The Hue-man, In Form and Function

DrYogi
Hari Simran Singh Khalsa, D.C

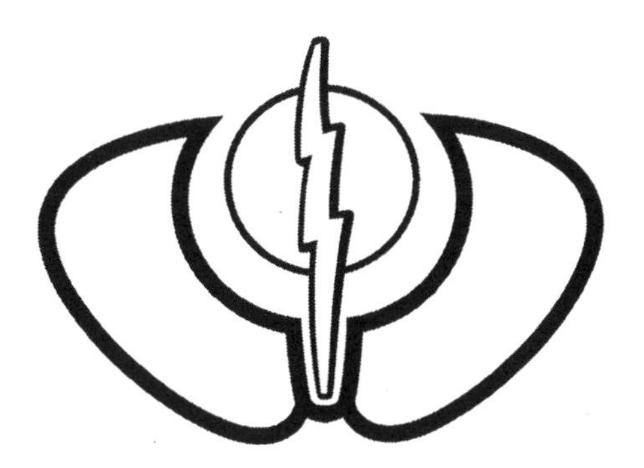

Author: Hari Simran Singh Khalsa, D.C./DrYogi
Editor-in-Chief: Sat Rattan Kaur Khalsa
Assistant Editors: Guru Sahai Kaur, Alexis, Shant Kaur Khalsa, Rudee Roth, Elaine Esparanza, and Sahib-Amar Kaur Khalsa
Special thanks to: Guru Raj Kaur Khalsa, Gurudass Singh Khalsa, Gurucharan Singh Khalsa, and Satya Singh Khalsa
Layout and Design: Lila Baker; Nick Baker, Fotoblender Services
Illustrator: Aurora Crispin
Illustrator: Justin Dean Hovarter
Assistant Illustrator: Sat Rattan Kaur Khalsa
Cover Photograph: Sat Rattan Kaur Khalsa
Model: Alexis
Publisher: Golden Dragon Productions www.DrYogi.com

The suggestions in this material in no way are intended to replace the advice of your healthcare professional. The diet, exercise, and lifestyle suggestions in this book come from ancient yogi traditions. Nothing in this book should be construed as medical advice. Any recipes mentioned herein may contain potent herbs, botanicals and naturally occurring ingredients, which have traditionally been used to support the structure and function of the human body. Always check with your personal physician or licensed health care practitioner before making any significant modifications in your diet or lifestyle, to insure that the ingredients or lifestyle changes are appropriate for your personal health conditions and consistent with any medications you may be taking. The conditions and diseases included in this book are intended for your personal information, not for treatment purposes. Neither the author nor the publisher shall be liable or responsible for any loss, injury or damage allegedly arising from any information or suggestions in this book.

Dedicated to Yogi Bhajan

Contents

Foreword

Sat Nam. This new book, The Hue-man, In Form & Function by Dr. Yogi/Dr. Hari Simran Singh Khalsa is the book that I have been waiting for a long time.
I was involved with the creation of the Kundalini Yoga as taught by Yogi Bhajan textbook for teachers in training called The Aquarian Teacher. It was a formidable task in that a group of international teachers had to take thirty years of Yogi Bhajans' teachings and distill them into a cogent and systematic course of study for Level One teachers in training. It took almost a decade of meetings and hard work by many individuals to bring the course and materials into being. Now many thousands of teachers are being trained using the textbook and take for granted all those teachings that are in one place.
I feel the same about this book. I have pored through many anatomy and physiology textbooks over the years and I have not found one that has the balance in application that this fine book contains. It is simple, clear and has just enough detail about the systems of the body to give a solid practical working knowledge of anatomy and physiology for a student and teacher of Kundalini Yoga. In addition, the information, tips, and recommendations about the actual practice of Kundalini Yoga as taught by Yogi Bhajan are invaluable. This book is the product of many years of Dr Hari Simran Singh Khalsa's practical experience as a Chiropractor, College Level anatomy and physiology instructor, and Teacher Trainer. His sobriquet of DrYogi is very apt; he brings the East and West together seamlessly. I hope that you use this book and enjoy it as much I have.

Kind regards and blessings,
Nirvair Singh Khalsa
CEO Kundalini Research Institute

Acknowledgments

First, I would like to acknowledge my parents, Robert and Helen Lee. They always inspired me to be persistent in the pursuit of excellence and happiness.
Thanks to the many assistant editors including Guru Sahai Kaur, Shant Kaur Khalsa, Alexis, Larry, Rudee Roth, Sunder Kaur, Elaine Esparanza, and Sahib-Amar Kaur Khalsa for their constant faith in the project.
Thank you to both Lila and Nick Baker for formatting this book with care and creativity.
Many thanks to our artists, Aurora Crispin and Justin Dean Hovarter
Special thanks to my lovely wife and Editor-in-Chief, Sat Rattan Kaur and my chiropractor, Dr. Shawn Dill.
Thanks to my dear friends Dr Fred, Sat Daya Singh, Ray, Tish and Marty, Japa Kaur, Michael, Julie Simran Singh, Siri Gian Kaur, Sukhbir, Parmatma Kaur, Snatam Kaur, Sevak Singh, Baba Siri Singh and Robin, Dev Murti Singh and Hari Amrit Kaur, Jeffery, Gurukar Singh, Chris, Shakta Kaur and Hari Dev Singh, and our friends at Life West: Dr Bruce Chester, Dr Kim Khauv, Dr Shakti Singh Khalsa, Dr Gerry Clum, Dr Brain Kelly, Dr Debbie Lindemann, John Boss, Phd., and Sue Ray.
Thanks to Yogi Bhajan for the inspiration to keep up.
Thank You ALL for the support and the endless encouragement to finish this project.

Teaching this Material

Anatomy & Physiology for Yogis

The goal of this material is to provide Kundalini Yoga teachers and Kundalini Yoga practitioners with basic knowledge of the structure and functions of the human body.

Objectives

- To develop the relationship between knowledge (anatomy and physiology) and the practice of yoga
- To provide insight into the relationship between structure and function
- To increase the vocabulary of the Kundalini Yoga teacher so that discussions with health professionals may be more effective

Utilizing this Material

Use yoga experiences to relate functions with anatomy. Using visual aids, such as models, videos and charts, can be of great value. Having a health professional who also teaches Kundalini Yoga is suggested, although the material is designed for anyone to teach. The study points are designed to aid in the review of the material.

Teachers are not Doctors

Despite the amount of anatomy and physiology knowledge that we, as teachers, may acquire we are usually not doctors. Only specially licensed healthcare professionals have the right to develop a professional healthcare relationship with a patient. Unless you are a doctor, any comment you make may expose you to the liability of practicing medicine without a license. There are two things to remember regarding this, the student-teacher relationship, and the subject of health.

1. You can always share experiences and ancient knowledge: "Garlic has for thousands of years been known by ancient Yogis to help boost the immune system," or, "I have noticed that when I have a cold and I eat garlic, I seem to feel better, sooner."
2. Any knowledge you share should always include a statement like: "I suggest seeing your Healthcare Professional about any health condition or health concerns."

While we're on the subject, please note the following:

The suggestions in this material in no way are intended to replace the advice of your healthcare professional. The conditions and diseases included in this book are intended for your personal information, not for treatment purposes.

INTRODUCTION

Everything inside is outside,
As above, so it is below,
The inside and the outside are all the same.

The study of functional anatomy provides numerous insights into the physiology of Kundalini Yoga. The human body is a sophisticated bioelectrical mechanism of infinite cosmic possibilities.

The body is often viewed as skin or the outer shell. The inner-workings of the human body include a fascinating series of events, which occur in rhythm and synchronization to create a temporary home for your Soul. The human body is designed to provide experiences that give us a feel for the nature of the universe, as it is, and as it might be.

Every action of the human body involves a symphony of contractions, secretions, electrical charges, and tissue changes. We can think of the brain as the command center or emperor of our physical universe. The systems are interdependent city-states. The body stays in balance (homeostasis) due to the coordination of constant electro-chemical input and output (or communication) between systems.

Disease is malfunction caused by an interruption in the input or output of information. A system malfunctions in three basic ways: It secretes too much, too little, or the wrong kind of material.

The amount of innate intelligence present in every one of our 100 trillion cells is no less than a reflection of the totality of the grandeur and awe of the entire universe. Beauty, majesty, humility, and profound excellence are qualities essential to the biological nature of the human organism. Human beings on the other hand, can walk, talk, chew gum, and behold the sacredness of a bird in flight all at the same time. The basic responses of the body operate without your conscious awareness while you are awake, asleep, or at the movies. You stay alive automatically. The complexity of the human organism is truly vast and involves a soup of neuro-chemicals.

Beyond structure and function is the realm of experience. The heart is where we find our compassion, passion, and taste for life. The yogi tastes life through the experiences of the elements (tattvas): Earth, water, air, fire, and ether. Humans experience the elements through the sensations.

Earth is felt through the yogi's relationship with gravity, the sense of smell, and the experience of being very grounded at the first chakra. Elimination and the ability to let go are managed by the earth element.

Water controls the emotions, is experienced through the sense of taste, and has a consciousness of the qualities of water. Water likes to be with more water, it likes to flow; it has density and can nurture the nature of passion. Water manifests its creativity at the second chakra. The human body is about 80% water, and is affected by the moon's gravity.

The Fire element changes all other elements, to the next element. Fire is the great transformer.

Fire controls digestion and is experienced through the sense of sight. There are three fires in the human body: the fire of digestion at the navel point (third chakra), the fire of respiration at the heart, and the fiery passion of reproduction at the gonads. Fire is at the core of metabolism and general good health.

Air is the element of life. The element of air is experienced through the sense of touch as when we feel the wind across our skin and in our lungs. The flow of air is related to the fourth chakra, the heart center. Air is a form of our basic energy called Prana.

Ether is the space between the air. It is experienced through the sense of hearing. We use ether to speak from the throat, which is the home of the fifth chakra.

All the elements are somewhat contained in each other. Understanding the sensitive relationship between the elements (tattvas) and the chakras is the core of understanding the physiology of Kundalini Yoga. Becoming sensitive to the flow of elements provides infinite insight into the individual self, the unlimited self, and the vast expanding universe.

The magnificent design of the human body is a masterpiece of functional divinity.

When the body is gone and your mind has left, the soul will still prevail. As soon as we are born, we prepare to die. May you dwell in God, and may your limited ego and ignorance die before you do! Be positive. Live long and prosper.

Sat Nam
&
Wahe Guru
Dr Hari Simran Singh Khalsa/DrYogi

Chapter One

Cells

The cell is the basic unit of life. Some live as single cells, and some are members of specialized tissue teams, gathered for a single purpose. The human body team has an estimated 50 Trillion Cells. Most members of this team begin as undifferentiated cells, and then adapt to the specific task, developing specialized characteristics necessary for their function. Cells can respond to the environment, learn to find food and reproduce. Cells are characterized by the ability to reproduce, grow, metabolize, excrete waste, move and respond to environmental stimuli.

Cellular Tensegrity

The Cellular Tensegrity Model proposes that the whole cell has internal prestressed structure. This cytoskeleton comprises microtubules and filaments that stiffen the cells to transfer forces both within the cell and within the entire tissue system. Tensegrity Models transfer mechanical forces, develop electrical fields, and facilitate communication within the system. This may be the way individual cells work together, communicate and adapt.

Basic Structure/Function

Cells were discovered by Robert Hooke in 1665. The word "cell" comes from the Latin word cellula, which means, "a small room." By weight, cells are 80% water, 15% protein, 3% fat, 1% carbohydrate, and 1% nucleic acids and minerals.

Cell Membrane

Surface specializations

The cell membrane is the plasma outer covering of the cell that keeps the cytoplasm in (cell water and internal organs), and the environment selectively out. Cell to cell communication and responses to the environment are mediated by the cell membrane. Some researchers have called the membrane "the brain of the cell." The cell membrane is made primarily of fat (lipid) with specialized embedded proteins. These proteins control the gateways in and out of the cell and receive chemical and electrical signals from the environment and other cells.
Many cell types demonstrate specialized surface modifications. Some cell types increase their surface areas, others develop special communication functions and some alter their shape, growing cilia for motility or projections for attachment.
Cells communicate through a variety of methods.

Cytoskeleton

The cytoskeleton is comprised of fibers and tubes that make the structural foundation of the cell.

Cytoplasm

The cytoplasm is the space occupying fluid that is inside the cell membrane. It contains the membrane bound organelles and the cytoskeleton.

The Nucleus

Nucleus comes from the Latin word for *nux*, which means nut. It contains the blueprint for reproduction and protein synthesis. The nucleus is the home of DNA and RNA.

Nuclear Envelope

The nuclear envelope is a doubled-layered membrane that encloses the contents of the nucleus and regulates the transport of genetic material in and out of the nucleus.

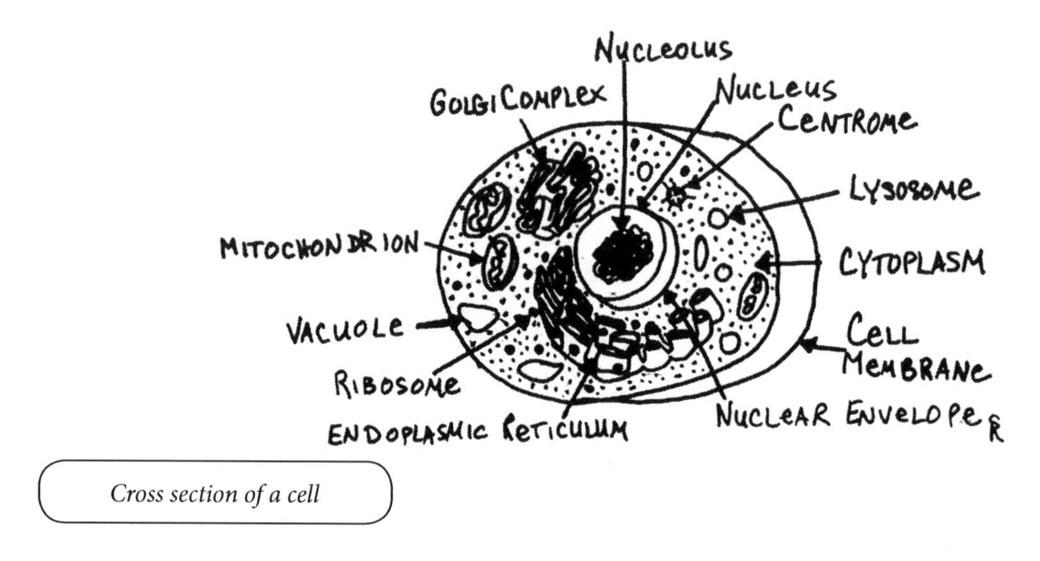

Cross section of a cell

The Nucleolus

The nucleolus is inside the nucleus. It consists of densely packed chromosome regions together with some protein and some RNA strands. The nucleolus initiates the formation of ribosomes.

Ribosomes

Amino acids are strung together for protein synthesis. The design is from a copy of the original DNA blueprint (RNA), provided by the nucleus.

Endoplasmic reticulum

Protein transportation and synthesis of more mature proteins and molecules and performs several chemical functions. The endoplasmic reticulum can store Ca+ in muscle cells and break down toxins in liver cells.

Golgi complex

The Golgi complex is a stack of fluid filled membranes that package proteins for transport either locally or for export. It is named after Camillo Golgi, an Italian Physician, who identified it in 1898.

Mitochondrion

The mitochondrion is the powerhouse for the cell, where raw materials are converted to energy by adding oxygen.

Lysosome

Lysosomes are membrane bound sacs of enzymes that have a great capacity to breakdown waste and damaged cell parts.

Vacuole

Vacuoles are membrane lined transport containers that surround a mass of fluid. In that fluid is nutrients or waste products.

Centrosome

The centrosome is a small body adjacent to the nucleus, which contains centrioles and organizes microtubules.

Chapter Two

Nervous System

Basic Structure/Function

One of the first structures to develop in the human embryo is the notocord, which serves as the nervous system, coordinating all actions and functions of development and growth. From your nose to your toes, and from conception to grave, your nervous system synchronizes your life.

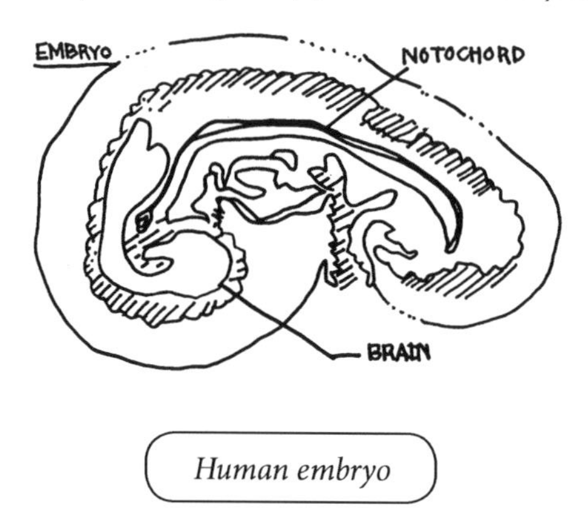

Human embryo

The Nervous System is responsible for sending and receiving communication between the brain and all the other cells, tissues and organs. The brain orchestrates and organizes the areas of function. Cell to cell, tissue to tissue and organ to organ, the nervous system coordinates and orchestrates the many functions of the body. The nervous system is the hardwiring of the body, taking messages from the brain to organs, muscles, glands, etc., as well as conveying messages from the body back to the brain from sensory neurons, postures and movement. This auto regulation is often referred to as homeostasis. When the signals from the nervous system are interrupted, the body looses its coordinated sensory-response system.

Neurons/Nerve Cells

Neurons are the structural and functional units of the nervous system. They are specialized to carry electrochemical signals to and from different areas of the nervous system as well as between the nervous system and other tissues and organs. Different types of nerve cells exist that respond to touch, light, sound and other stimuli. Some nerve cells are several feet long and some are very short.

Neurons are divided into four categories:

1. Cranial nerves: Connects the senses of hearing, seeing, smelling and tasting to the brain
2. Central nerves: Connects the brain to the spinal cord
3. Peripheral nerves: Connects the brain to the extremities
4. Autonomic nerves: Connects the brain and spinal cord to the organs of the body

The basic structure of a neuron helps it carry out its function.

Soma: The Soma is often called the cell body. It is the central part of the neuron that contains the nucleus and produces all the proteins for the dendrites, axons and axon terminals. It also contains specialized organelles.
Nucleus: The nucleus contains chromosomes, which consist of molecules of DNA that hold the information for cell development and synthesis of proteins necessary for cell maintenance and survival. As DNA cannot pass through the nuclear membrane, the nucleus synthesizes RNA from DNA and ships it through its pores to the cytoplasm for use in protein synthesis.
Dendrites: Dendrites surround the cell body and receive signals from other neurons or cells. This electrical charge is transferred through the cell body onto the axon.
Axon: The Axon is a long thin extension of a nerve cell, which is specialized for conducting signals from one nerve cell to another. The nerve signals travel along the axon away from the cell body and toward synapses at the axon terminals. Each nerve cell has only one axon. It is covered by a myelin sheath, which acts as insulation and increases the transmission speed along the axon.
Axon terminals: At the other end of the axon, the axon terminals transmit the electro-chemical signal across a synapse (the gap of about 3 nanometers between the axon terminal and the receiving cell). Within this "gap" is a bioelectrical event, which stimulates the nerve to conduct the neural message along its path to the brain or from the brain. Any nerve tissue that is damaged is replaced by scar tissue. This scar tissue can interrupt the message transmission.

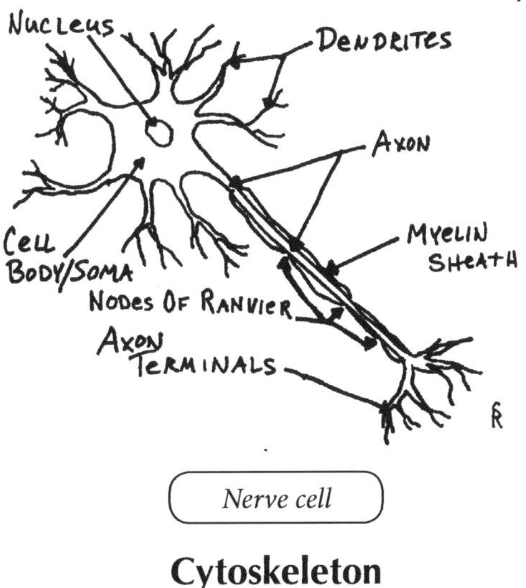

Nerve cell

Cytoskeleton

Inside the nerve is a structural cytoskeleton composed of microtubules. These microtubules are arranged into filaments, which arrange themselves in an imperfect helix containing 13 tubulin dimers. Researchers have theorized that this structure maybe the architecture and mechanism that facilitates cellular tensegrity and consciousness.

Support Cells

Support cells such as, Schwann cells, and glial cells (oligodendroglia, astroglia, and microglia), maintain homeostasis, form myelin, provide physical and nutritional support for neurons, insulate one neuron from another, destroy pathogens, digest parts of dead neurons, and modulate neurotransmission. These cells do not transmit electro-chemical signals.

Myelin Sheaths

The nervous system has a fatty covering of myelin. It acts as an insulation to isolate the nerve tissue. Sometimes when this myelin sheath is damaged or the tissue is irritated, you will notice a pattern of shaking that exists. If the nerve is allowed to keep firing, the nerve will recruit more sheathing material, creating a more stable nerve junction, lessening the sensation of shaking. Some diseases, however, disrupt the nerve transmission permanently and progressively.

Nodes of Ranvier

The Nodes of Ranvier are the gaps formed between the myelin sheaths along the length of the axon. These nodes contain channels for sodium ions, which boost the electrical charge required to pass signals from one nerve to another.

Organs in This System

The nervous system is divided into the Central Nervous System and the Peripheral Nervous System.

A. Central Nervous System (CNS)

The CNS consists of:

- The Brain
- The Brain Stem
- The Spinal Cord

B. Peripheral Nervous System (PNS)

The PNS consists of:

- Sensory Neurons Afferents: Nerve receptors that inform the Central Nervous System of sensory information.
- Motor Neurons Efferents: Nerves that send a signal to specific tissues like muscles, glands and organs.

The PNS is subdivided into:

- Sensory-Somatic Nervous System.
- Autonomic Nervous System.

Central Nervous System

The Brain

The brain is a three-pound organ of the central nervous system, with the consistency of soft gelatin or soft tofu, located in the cranium. It is the central controlling and regulating mechanism that manages the many functions of the body. The brain is the headquarters and command center of the body. It is very complex and contains approximately 50 to 100 billion neurons, which connect to each other through approximately 1000 trillion synapses. Each cubic millimeter of cerebral cortex contains approximately 1 billion neurons. The brain processes.

The brain provides central control so the human can adapt to the environment, situation and opportunity. It accepts approximately 1 million pieces of information per second, but can only directly manage approximately 200 pieces of information per second. The ability to manage the brain, the mind, the images, and the flow of thoughts is an important aspect of the practice of Kundalini Yoga, the science of success, and your ability to be happy.

The Brain has many functions.

Memory - Who, what, why, when, etc.,
Forgetting - Letting go of memories.
Somatic (Body) functions - Heartbeat, breathing, digestion
Voluntary and involuntary movement - Muscle motion, breathing, talking
Physiological adaptation - Temperature, light, blood pressure

The Brain consists of:

Cerebrum: It consists of the frontal, parietal, temporal and occipital lobes. There are two hemispheres separated by a huge band of fibers called the corpus callosum, which transmits messages between them.
Frontal lobe: The frontal lobe is the part of each hemisphere of the brain located behind the forehead that serves to regulate and mediate the higher intellectual functions. This controls the personality. It is associated with intelligence, memory, judgment, and motor control.
Parietal lobe: The parietal lobe sits between the occipital lobe above it and behind the frontal lobe. The parietal lobe is divided into right and left parietal lobes that correspond to the right and left brain hemispheres. The parietal lobes have an important role in integrating our senses. In most people, the left side parietal lobe is thought of as dominant because of the way it structures information to allow us to make calculations, read and write, produce language, and perceive objects normally. The non-dominant lobe receives information from the occipital lobe and helps provide us with a 'picture' of the world around us.
Temporal lobe: The temporal lobes are located on the bottom and at the side of each of the two brain hemispheres. They are concerned with perception and recognition of auditory stimuli, important for the processing of semantics in speech and vision. It contains the hippocampus and plays a key role in the formation of long-term memory.
Occipital lobe: The region in the back of the brain, which processes visual information.
Cerebellum: It controls coordination of muscular activity and patterns of movement, muscle tone, reflex actions, and equilibrium. Its left and right cortex divide key functions. It is related to memory and a core sense of awareness, proprioception and posture.
Pons: The portion of the brain lying above the medulla oblongata and below the cerebellum and the cavity of the fourth ventricle. It consists of a thick bundle of nerve fibers that links the medulla oblongata to the thalamus. The pons is the synaptic relay station between the cerebrum and cerebellum coordinating hearing, taste, eye movement, rhythm, and depth of breathing.
Medulla oblongata: It is the lower portion of the brainstem. The medulla oblongata contains reflex control centers and is the exit point for the cranial nerves. This part of the brain directly controls breathing, blood flow, swallowing, and other essential functions.

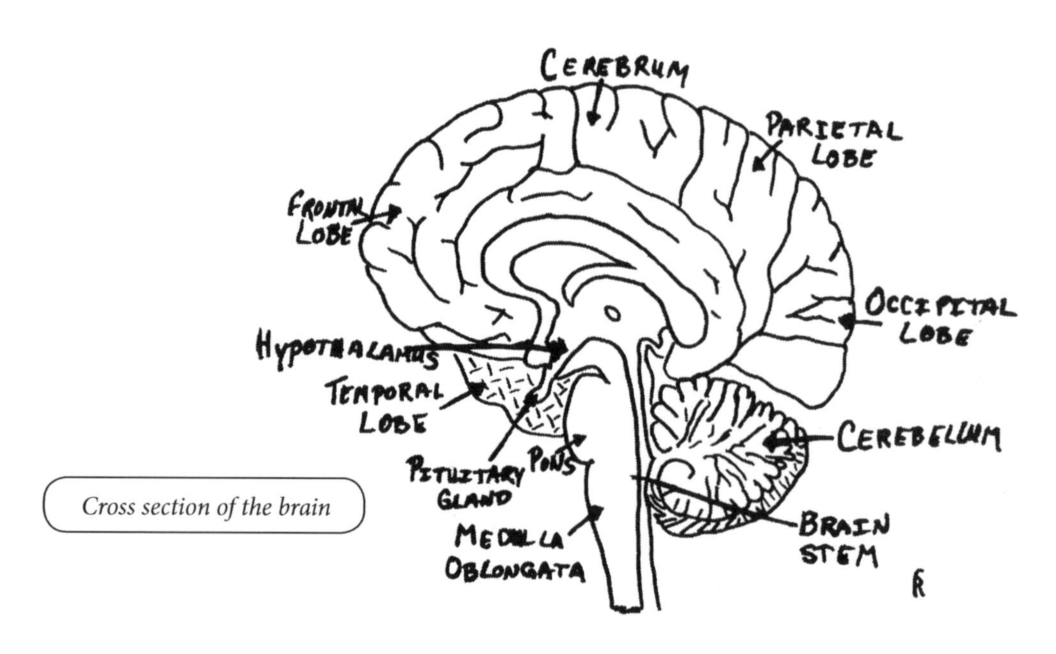

Cross section of the brain

Hypothalamus: It is a region of the brain, between the thalamus and the midbrain, that functions as the main junction box of the autonomic nervous system. The hypothalamus regulates sleep cycles, body temperature, appetite, moods, sex drive, thirst, and the release of hormones from many glands, especially the pituitary gland. Parasympathetic functions appear to be directed by the medial and anterior hypothalamic regions, whereas lateral and posterior areas contain sympathetic centers. Both centers put forth their effects by way of relays through brain stem centers.

Optic Chiasm

The optic nerve crosses at the optic chiasm. This crossover point is also the location of the pituitary and hypothalamus gland. The position of our eyes has multiple effects on our brain. This is how the placement of the eyes affects the secretion of the nervous system and glandular system through the pituitary gland. The pituitary is where the nervous system and the glandular system meet, communicate, and centrally control the neuro-glandular systems.

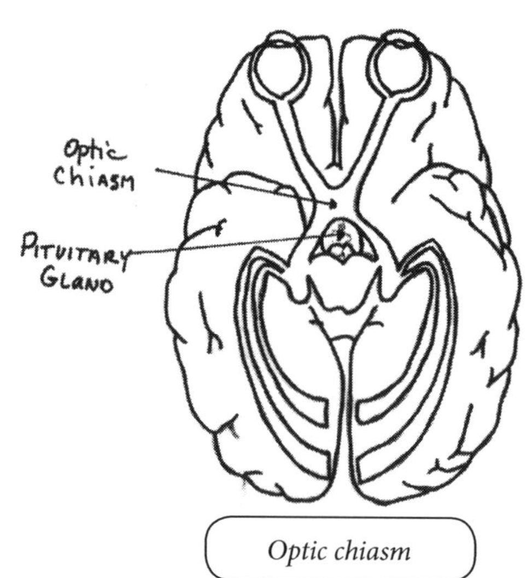

Optic chiasm

Blood Brain Barrier

The brain has specific needs and operates in a specialized environment. The capillaries within the Central Nervous System have a special type of membrane called the blood brain barrier. This membrane does not allow the same degree of permeability as the rest of the Cardiovascular System. The blood brain barrier has tight junctions, which allow glucose, oxygen, Co2, hormones and a few other small molecules in, yet keep bacteria, other larger molecules, and other Trojan horses out of the cerebrospinal fluid (CSF) and away from the sensitive nerve tissues of the brain. The barrier is a thick fibrous membrane and has astrocytic endfeet. Astrocytic endfeet are star shaped glial cells, which release neurotransmitters, called gliotransmitters, that interface with the Nervous System (CNS & PNS or just nervous sysem in general). Gliotransmitters play an important role in tissue repair following traumatic injury.

The Brain Stem

ALL the nerves from the brain form a bundle (the brainstem) and pass through the foramen magnum and the first 2 vertebras(C1 and C2); the most vulnerable region of the spine to trauma and injury.

Any interruption in the neural signal from the brain will cause malfunction, pathology or maladaptation in the body. Correct neck lock allows the brainstem to be in an optimal position to facilitate communication between the cells of the brain and the cells of the body. If correct neck lock is utilized during Kundalini yoga, amazing postural changes can occur. This is due to the amount of proprioceptive mechanoreceptors located in the upper cervical spine.

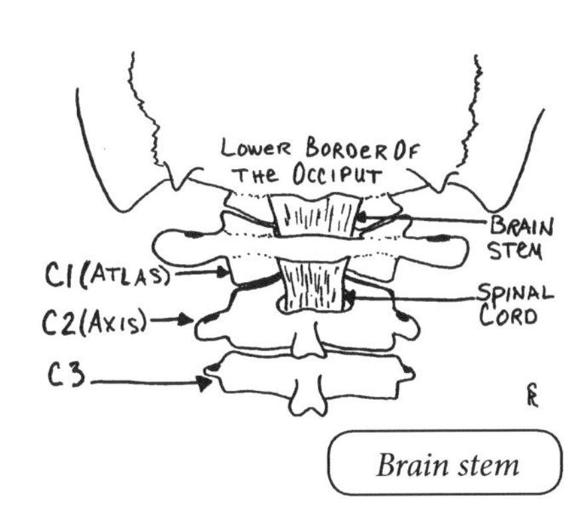

Brain stem

The Spinal Cord

The spinal cord is a thick bundle of nerve tissue from the Central Nervous System. It extends from the brain stem (medulla oblongata), through a large hole (foramen magnum) in the base of the skull (occiput), and down through the spinal column to the second lumbar vertebra. The spinal cord is about 18 inches long.
The terminal portion of the spinal cord is called the conus medullaris.

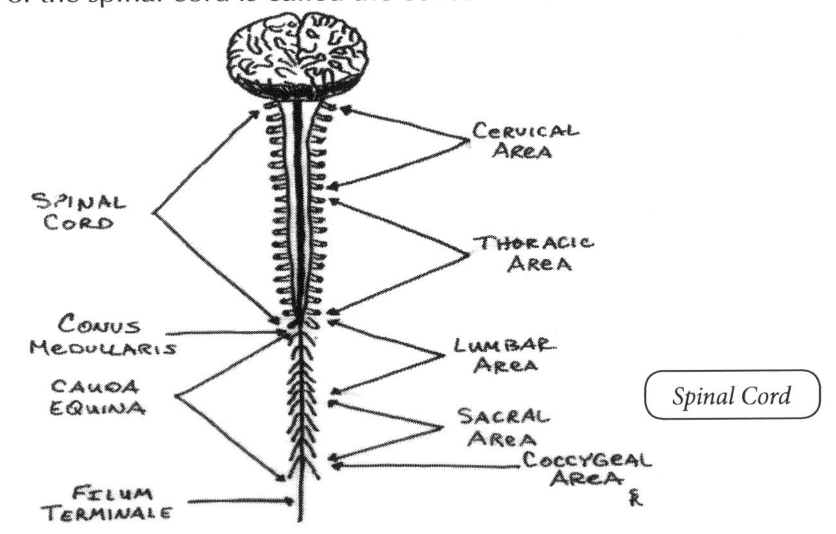

Spinal Cord

Below the conus medullaris, the spinal nerves continue as dangling nerve roots called the cauda equina ("horse's tail") that travel through the vertebral column. The cauda equina forms because the spinal cord stops growing in length around the age of four, even though the vertebral column continues to lengthen until adulthood. As a result, the sacral spinal nerves originate in the upper lumbar region.

The pia mater continues as an extension of the cauda equina and is called the filum terminale. It anchors the spinal cord to the coccyx.

At the level of each vertebra in the spine, nerve fibers arise from the spinal cord and emerge through openings between the vertebrae (intervertebral foramina). These nerve fibers are the spinal nerve roots. The Peripheral Nervous System carries messages to and from the brain.

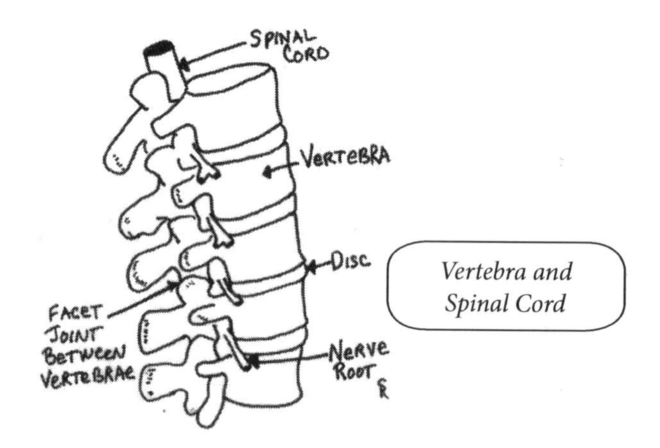

Vertebra and Spinal Cord

Meninges: The meninges are three layers of membranes, which completely surround the brain and spinal cord. This system of membranes are continuous. The meninges and the cerebospinal fluid protect the Central Nervous System.

Pia mater: The thin delicate membrane that covers the entire spinal cord.

Dura mater: A tough covering that creates a sac around the spinal cord, pea mater, and arachnoid mater.

Arachnoid mater: The arachnoid (arachnid = spider) resembles a spider web that connects the spinal column (the bones) to the spinal cord (main nerve channel). It is the inner membrane and it contains cerebrospinal fluid. The arachnoid membrane has an important function – keeping the spinal cord properly positioned in the spinal column. Any deformation, twist or tension on the arachnoid affects the nervous system and spine.

Subarachnoid: The space between the spinal cord and the arachnoid mater. It is full of cerebrospinal fluid (CSF).
The dura mater is attached to the skull, and to the bones of the vertebral canal in the spinal column. The arachnoid is attached to the dura mater, and the pia mater is attached to the spinal cord.
The spinal cord has 31 pairs of nerve roots that exit laterally along the spine, controlling somatic (body) functions. Each nerve services a specific region of the body. Any impedance to the nerve flow will affect all three functions: sensory, motor and organ function.

Cerebrospinal Fluid

Cerebrospinal fluid (CSF) = brain soup; this fluid not only cushions and supports the spinal column but also circulates the neurochemical messages. The choroid plexus secretes about 500 mL of cerebrospinal fluid per day. The composition of CSF is similar to an ultra filtrate of blood plasma.
Cerebrospinal Fluid Dynamics:

There are three wave motions associated with the spinal fluid:
The first wave is that associated with the breath. Each time there is a breath, the sacrum moves and the spinal fluid creates a wave like the tides on the ocean.
The second wave occurs from the constant polarizing and depolarizing of the hemispheres of the brain. This is the mental wave and is like the currents of the ocean.
The third wave is of unknown origin and is referred to as the primal wave. This wave is much more subtle and delicate. It coordinates the other two waves by aligning their peaks and

valleys in the most advantageous, harmonious rhythm.
Many osteopaths have written about this primal wave, namely Sutherland and Korr. The mind and the breath are linked through the primal wave.

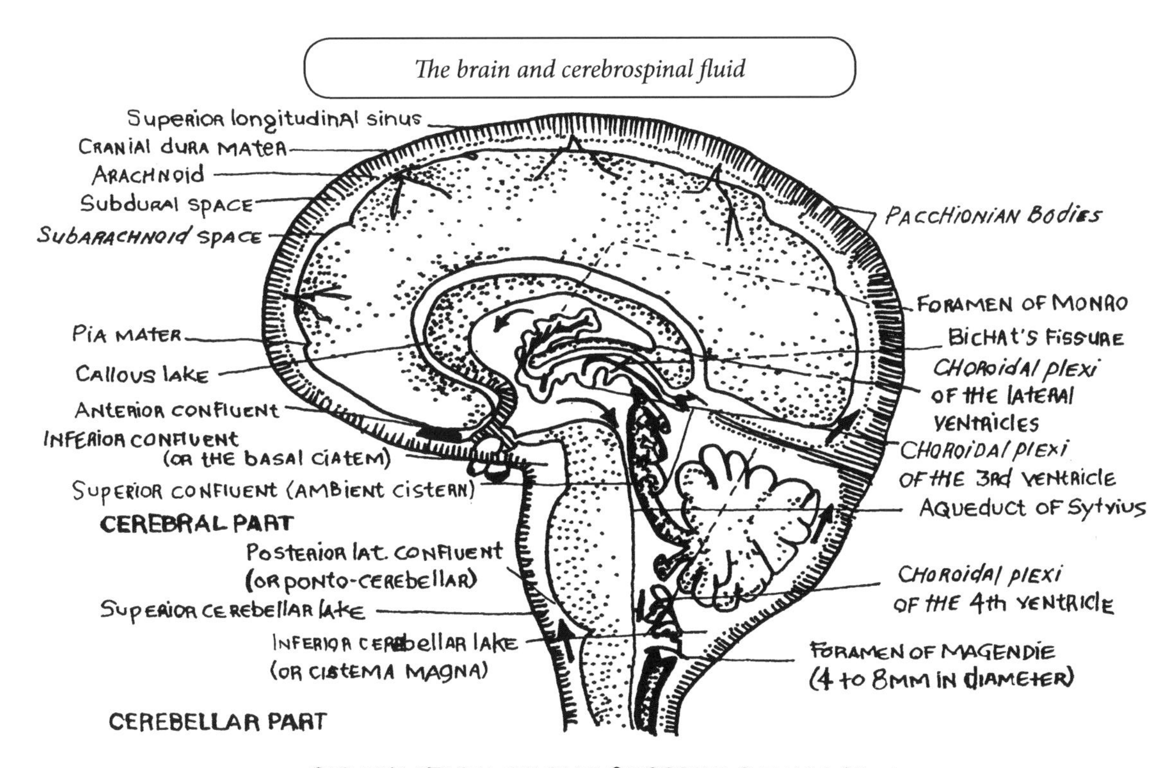

The brain and cerebrospinal fluid

Cerebrospinal Fluid Flow					
COMPONENT	CSF**	SERUM**	COMPONENT	CSF**	SERUM**
Water content (%)	99	93	K+ (meg/liter)	2.8	4.5
Protein (mg/dl)	35	7000	Ca++ (meg/liter)	2.1	4.8
Glucose (mg/dl)	60	90	Mg++ (meg/liter)	2.3	1.7
Osmolarity (MOsm/liter)	295	295	C1- (meg/liter)	119	102
Na+ (mg/liter)	138	138	Ph	7.33	7.41
**CFM and **SERUM are average or representative values. Source: adapted from Dr.Fishman(1980)					

Quality of Cerebrospinal Fluid

It takes forty bites of digested food to create one drop of blood. It takes forty drops of well-circulated blood to create one drop of semen. It takes forty drops of high quality semen to create one drop of cerebrospinal fluid.

Foods that are known to be very advantageous for healthy spinal fluid:

- Onion, ginger, garlic at 3:2:1 ratio;
- Yogi Tea, black sesame seeds, 14R Formula,
- Proper protein-carbohydrate fat balance, olives, walnuts, garlic, garlic, and garlic!
- Oranges help eliminate mental toxins.
- Herbs that are beneficial for CSF: gota kola and gingko.
- Brain hygiene: subconscious cleansing — Kirtan Kriya.
- Blood cleansing: Breath of Fire.
- Radiant health: long clean hair.

The Peripheral Nervous System

The Sensory-Somatic Nervous System

The Sensory-Somatic System consists of:
13 pairs of cranial nerves and
31 pairs of spinal nerves

Cranial Nerves

Cranial nerves are hardwired to the brain. They emerge directly from the brain and brain stem, carrying sensory and/or motor information. These messages are an important part of our awareness and have a direct effect on consciousness. There are twelve pairs plus one of cranial nerves.
Four sets of cranial nerves emerge from the cerebrum. The other cranial nerves emerge from the brainstem at the base of the skull and branch out to the face, throat, stomach and other parts of the body.

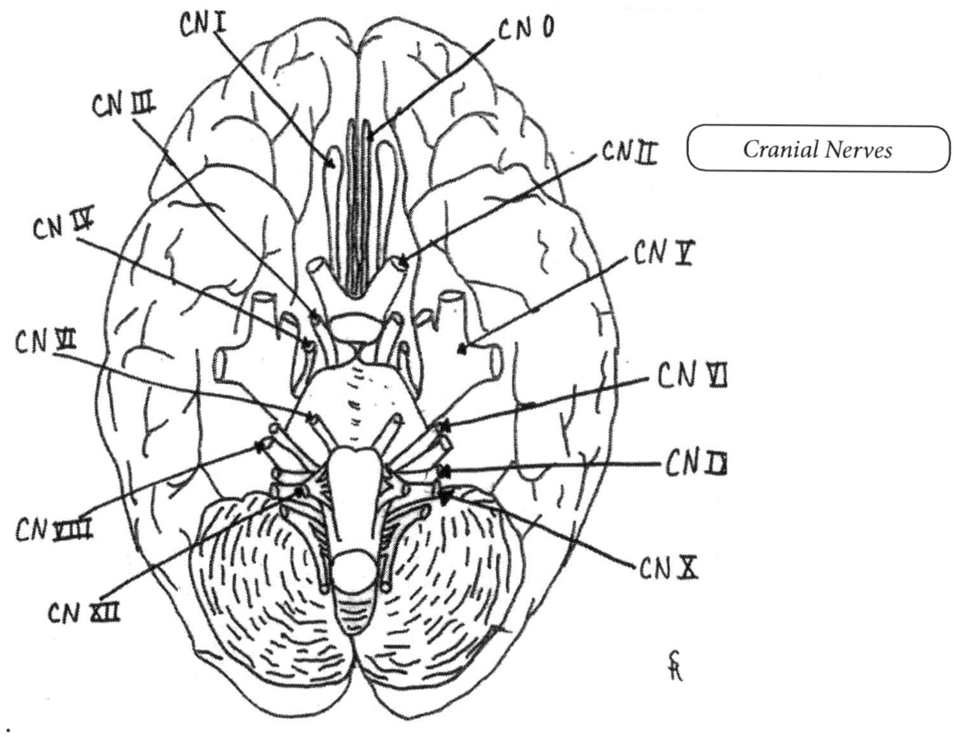

Cranial Nerves

Cranial Nerves		
CN 0: Cranial nerve zero	Sensory	Smells and senses pheromones – olfactory
CN I: Olfactory nerve	Sensory	Smells – olfactory
CN II: Optic nerve	Sensory	Sees – vision
CN III: Oculomotor nerve	Motor	Moves eyes and constricts pupils, accommodates (focuses the lens for near vision)
CN IV: Trochlear nerve	Motor	Moves eyes
CN V: Trigeminal nerve	Sensory and Motor	Chews and feels front of head (face)
CN VI: Abducens nerve	Motor	Moves eyes
CN VII: Facial nerve	Sensory and Motor	Moves the face, salivates (submandibular and sublingual glands), cries, tastes, receives small amount of sensation around the external ear
CN VIII: Vestibulocochlear nerve	Sensory	Hears, receives vestibular input that helps control balance
CN IX: Glossopharyngeal nerve	Sensory and Motor	Swallows, salivates, tastes, monitors carotid body and sinus, receives small amount of sensation around the external ear
CN X: Vagus nerve	Sensory and Motor	Swallows, lifts palate, talks, communicates to and from thoraco-abdominal viscera, tastes, receives small amount of sensation around the external ear
CN XI: Accessory nerve	Motor	Turns head, lifts shoulders
CN XII: Hypoglossal nerve	Motor	Moves tongue

Spinal Nerves

There are 31 nerve roots that exit the spinal cord to serve as major pathways of communication between the brain (Central Nervous System) and the body (Soma).

The spinal nerves are components of the Peripheral Nervous System. They form bundles of nerves that serve a common function. These nerve bundles form the nerve plexises, which we know as chakras. The Cervical Plexus, Brachail Plexus, Lumbar Plexus, Sacral Plexus, Solar Plexus and the Coccygeal Plexus, all serve a specific region and deliver a specific function.

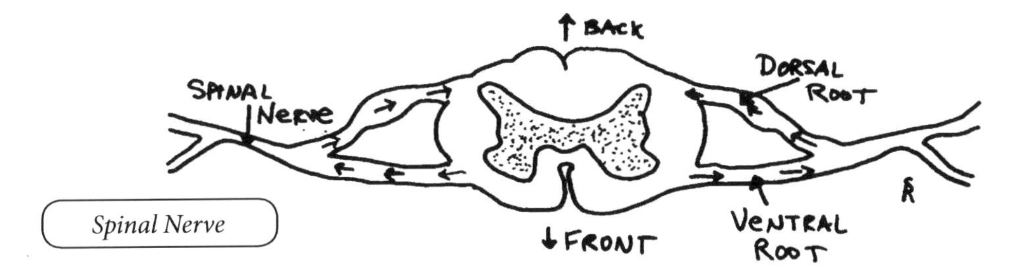

Spinal Nerve

Spinal nerves provide sensory information for the Central Nervous System and deliver nerve impulses from the brain to the muscle. Each spinal nerve has two roots. The ventral (front) root carries motor impulses from the brain and the dorsal (back) root carries sensory impulses to the brain.

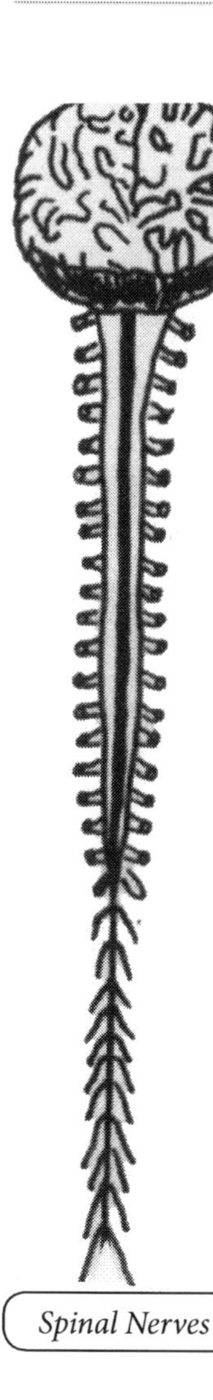
Spinal Nerves

The ventral and dorsal roots fuse together to form a spinal nerve, which travels down the spinal canal, alongside the cord, until it reaches its exit hole (intervertebral foramen). Each spinal nerve branches and each branch has both motor and sensory fibers. The smaller branch (posterior primary ramus) turns posteriorly to supply the skin and muscles of the back of the body. The larger branch (anterior primary ramus) turns anteriorly to supply the skin and muscles of the front of the body and forms most of the major nerves.

The spinal nerves are numbered according to the vertebrae above where it exits the spinal canal. The eight cervical spinal nerves are C1 through C8, the twelve thoracic spinal nerves are T1 through T12, the five lumbar spinal nerves are L1 through L5, and the five sacral spinal nerves are S1 through S5. There is one coccygeal nerve.

Each nerve carries sensory signals from sensory receptors (dermatome), nerves that control specific muscles (myotome), and nerves that facilitate organ function (visceral nerves).

If nerve fibers are damaged or impinged, the function of that nerve may be lost. When nerve impingement results from two vertebras being out of position, this is referred to as a subluxation. Organ malfunction, altered sensation, and muscle weakness can all be a result of a subluxation.

Sensory Neurons

There are several different types of sensory receptors:

Dermatomes

A dermatome is an area of skin innervated by a sensory nerve fiber. Dermatome maps show the typical pattern of what spinal nerve root feeds which part of the skin.

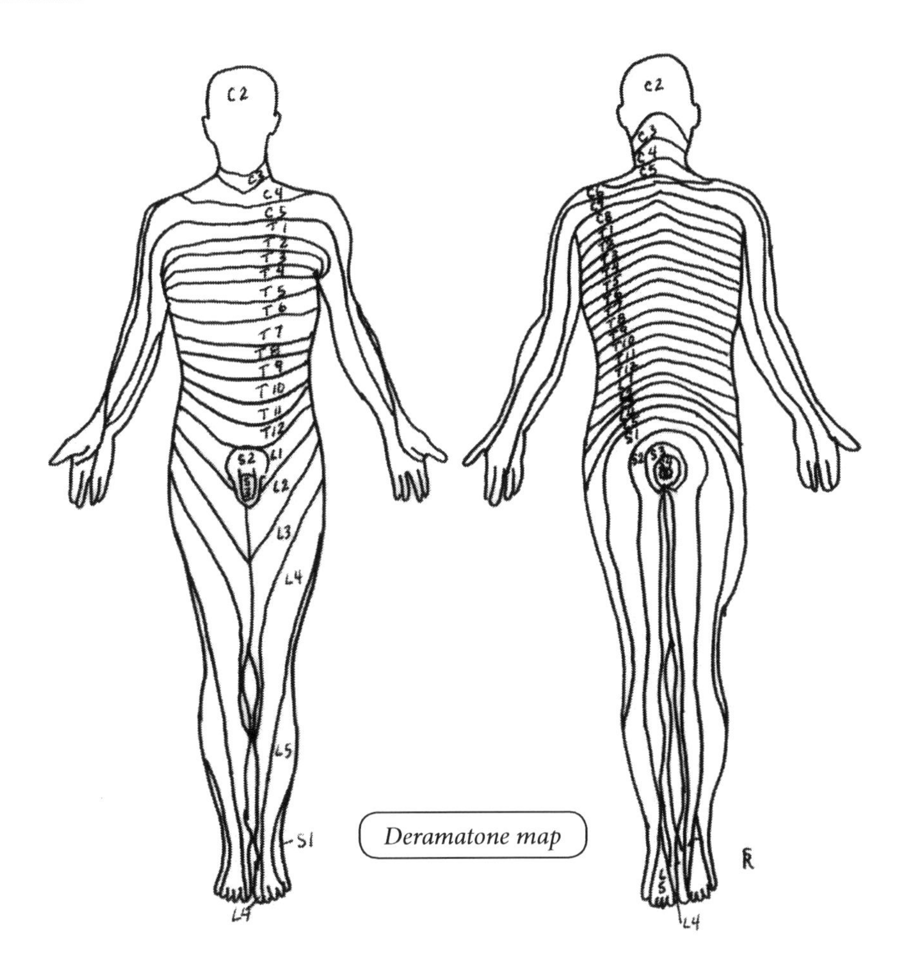

Deramatone map

Deramatomes	
C1	No sensory dermatome
C2	Posterior half of the skull cap
C3	Area correlating to a high collar
C4 & C5	Shoulders
C6 & T1	Inner and outer forearms
C6, C7, & C8	Thumbs and little fingers
T4	Inframammary fold
T6 & T7	Xiphoid process
T10	Belly button
L1	Iguinal ligament
L2	Fronts of thighs
L4	Knee, also provides sensation to the big toe
L5	Middle and sides of calves and three middle toes
S1	Ankle jerk reflex and little toes
S2 & S3	Genitalia

Myotomes

Each muscle in the body is supplied by specific segments of the spinal cord and by its corresponding spinal nerves. The muscle, and its nerve make up a myotome.

Myotomes	
C3, C4 & C5	Supplies the diaphragm
C5	Also supplies the shoulder muscles and the muscles that bend the elbow
C6	Bends the wrist back
C7	Straightens the elbow
C8	Bends the fingers
T1	Spreads the fingers
T1 –T12	Supplies the chest wall & abdominal muscles
L2	Bends the hip
L3	Straightens the knee
L4	Pulls the foot up
L5	Wiggles the toes
S1	Pulls the foot down
S3, S4 & S5	Supplies the bladder, bowel and sex organs and the anal and other pelvic muscles

Auditory Receptors

Motion and music stimulate hearing and balance. Inside the ear is a tube that is coiled and varies in diameter. Inside this cornucopia-shaped structure are small hairs, which are the auditory receptors. These receptors vibrate in different areas of the tube, sending signals of sound perception to the brain through the auditory cranial nerve. Inside this tube is also a liquid. The position of this liquid gives us our sense of orientation to gravity.

Aortic and Carotid Artery Baroreceptors

Baroreceptors are sensory receptors in blood vessels. They detect blood pressure and report abnormal blood pressure to the Central Nervous System. The brain innately responds by regulating the resistance of the blood vessels and the rate and strength of the heart's contractions. This process is the baroreflex. Baroreceptors work by detecting stretching in the blood vessel walls.

This is one important factor in avoiding injuries in the practice of Kundalini Yoga! When the practitioner rises up quickly and drops his/her head back, the baroreceptor located in the neck (carotid) can be stimulated; the blood pressure in the brain drops and often a loss of consciousness occurs. Do not wait until you hear the sound of thud!

If you are leading an exercise, alert fellow yogis to the possibility that they may feel a little lightheaded, and if they do, they should sit or lie down with care.

To avoid any potential serious head or neck injuries, be especially careful in places with very hard floors such as concrete, tile, terrazzo, and hard woods.

Examples:

- Moving from Guru Pranam into Camel Pose (the classic example)
- Triangle to Cobra or Camel Pose
- Miracle Bend
- Stand, sit down, stand up, sit down, stand up, and inhale deep and lean back

Nociceptors

Nociceptors are located throughout the body and respond to potentially damaging stimuli by sending signals to the spinal cord and brain. These signals are perceived as pain. There are thermal, mechanical, chemical and sleeping (silent) receptors. Sleeping receptors respond only to inflammation usually associated with trauma, infection or tissue injury.

Thermoreceptors

Thermoreceptors are the receptive portion of sensory neurons that senses hot and cold on the skin and separates internal temperature (hypothalamic thermostat).

Chemoreceptors

Sensory nerve cells, or sense organs, that respond to chemical stimuli.
Examples:

- Odor (olfactory neurons): cranial nerve 1.
- Blood oxygen, carbon dioxide, and hydrogen receptors: We commonly call this pH. These senses detect the effect of Breath of Fire (and all other Pranayama), eating a vegetarian diet, and also detect the challenges of eating meat and sugar, shallow breathing, stress and worry. The blood pH exists in a delicate balance with values between 7.41 and 7.37. A change of as little as .5 can cause tissue acidosis. The more acidic (smaller numbers) the blood becomes, the more the nervous system becomes agitated, irritated, and inflamed. The pH factor measures the amount of hydrogen in the blood and is a reflection of the efficiency of the body in utilizing oxygen (affected by long deep breathing) and food and how well the waste from the use of these fuels (Prana) can be eliminated (Apana). The Breath of Fire works directly on blood prana and apana balance.
- Blood glucose: Sensed in the hypothalamus, glucoreceptor.
- Blood osmolarity: Sensed in the hypothalamus, osmoreceptor.
- Taste: Sensed in the taste buds, located on the tongue.

Photoreceptors

Photoreceptors are located in the eye. The pineal gland is light sensitive as well, and so is our skin. The pineal gland is responsible for the timing of cycles in relation to the cosmic rhythm, like migration, the lunar cycle: menstruation, and all other thunderous cosmic rhythms of the universe.

Mechanoreceptors

A type of sensory reception that is subdivided into sensitivity to touch, pain, sound, gravity, muscle tone, and vibration. It is the basis for reflexes.

Motor Neurons

More Motion, More Health.

Accurate information into the system creates the best response out of the system. This is a common and continuous theme of both computer science as well as nervous system health. The sensors of the body provide the information for the Central Nervous System. This information allows the brain to develop an appropriate response to the environment. Proprioception is the unconscious perception of movement, location, and posture of our body in physical space. This information is gained by sensory receptors found all over our body as well as by the semicircular canals of the inner ear. The receptors constantly send messages through the Central Nervous System(CNS). The CNS then relays information to the rest of the body about how to react and with what amount of tension. Without proprioception, we would need to watch our feet consciously in order to make sure that we stay upright while walking. The practice of Kundalini Yoga is a poetic symphony of input for our sensory system - a juicy endocrine tango and ballet of neuro-chemicals that results in an integrated experience of synchronized ecstasy. The play of mechanoreceptors, Golgi tendon organs, and muscle spindles provides messages to the brain. These messages determine the "tone" of the nervous system and guide every aspect of our physiology.

Proprioceptive Messengers

Muscle Spindles

Scientists have noted that, next to the retina of the eye, the muscle spindle receptor apparatus is the most complex system within vertebrates. The muscle spindles are tiny fibers encapsulated in a fluid filled bag. There are two types of sensors in the muscle spindle: the nuclear bag fibers and the nuclear chain fibers. The bag fibers are a large number of nerve nuclei that congregate into a bag. There are 1-3 bags per muscle spindle. The nuclear chain fibers are about half the size of the bag fibers and are aligned in a long chain throughout the receptor area. Each spindle has 3-9 chain fibers.

There are sensors that specialize in static or dynamic positions. These sensors continuously send messages to the brain about position sense. These muscle spindle sensors have been found to be in dense concentration in the upper cervical region. Research has found that the sub occipital muscles have from 100 to 250 receptors per gram. In comparison, the trapezius and the gluteus maximums have 2.2 and .8 receptors per gram respectively.

Golgi Tendon

The Golgi Tendon organ(GTO) is a type of receptor that provides information about changes in muscle tension. It is activated by muscle and ligament tension. The GTO is located where the muscle and tendon meet and discharges from load or stretch. The GTO synapses in the spine facilitate lightning fast motion of both muscles and joints.

When the joints and muscles move well and send a complete message to the brain normal afferentation occurs. A subluxation(abnormal afferentation/signal to the brain) occurs when the joints of the spine do not move normally and the signal is altered. This type of dis-information is most common in the upper cervical spine due to the natural mechanics and the neurological implication. Global proprioception organs include: eyes, inner ear, skin sensors, joint mechanoreceptors, muscle spindles and Golgi tendon organs. Much of the proprioception occurs in the upper cervical spine.

Why Moving is Good for Your Health

As we move, the movement sensors (mechanoreceptors) send signals to the brain. There are many types of movement receptors including joint capsules and Golgi Tendons. The signal to the brain is called the afferent signal. The signal from the brain is called the efferent signal. There are about 36 afferents for every efferent signal.

This mechanism controls posture and balances the sympathetic and parasympathetic nervous system, which controls physiology. The signal (fine touch) first goes through the external cuneate nucleus located in the medulla oblongata (brain stem) and the central cervical nucleus. These signals then go to the cerebellum and to the postural reflexes.

The Intermediate Nucleus of the medulla is a site for the integration of cervical information and the generation of autonomic responses. The signal from the brain to the body keeps the parasympathetic nervous system (rest and digest) working well, and the sympathetic nervous system (fight or flight) ready to adapt to the information provided. When the signal is interrupted, the physiology does not receive the correct information. Excess inappropriate sympathetic response (sympatheticatonia) is stress, and is the cause of many illnesses.

The motions inherent to Kundalini yoga are exceptionally effective. Most of the movements are new to us and contain new rhythms. Doing new activities and having new experiences encourages the secretion of neural growth factor, a hormone that facilitates adaptation and personal growth.

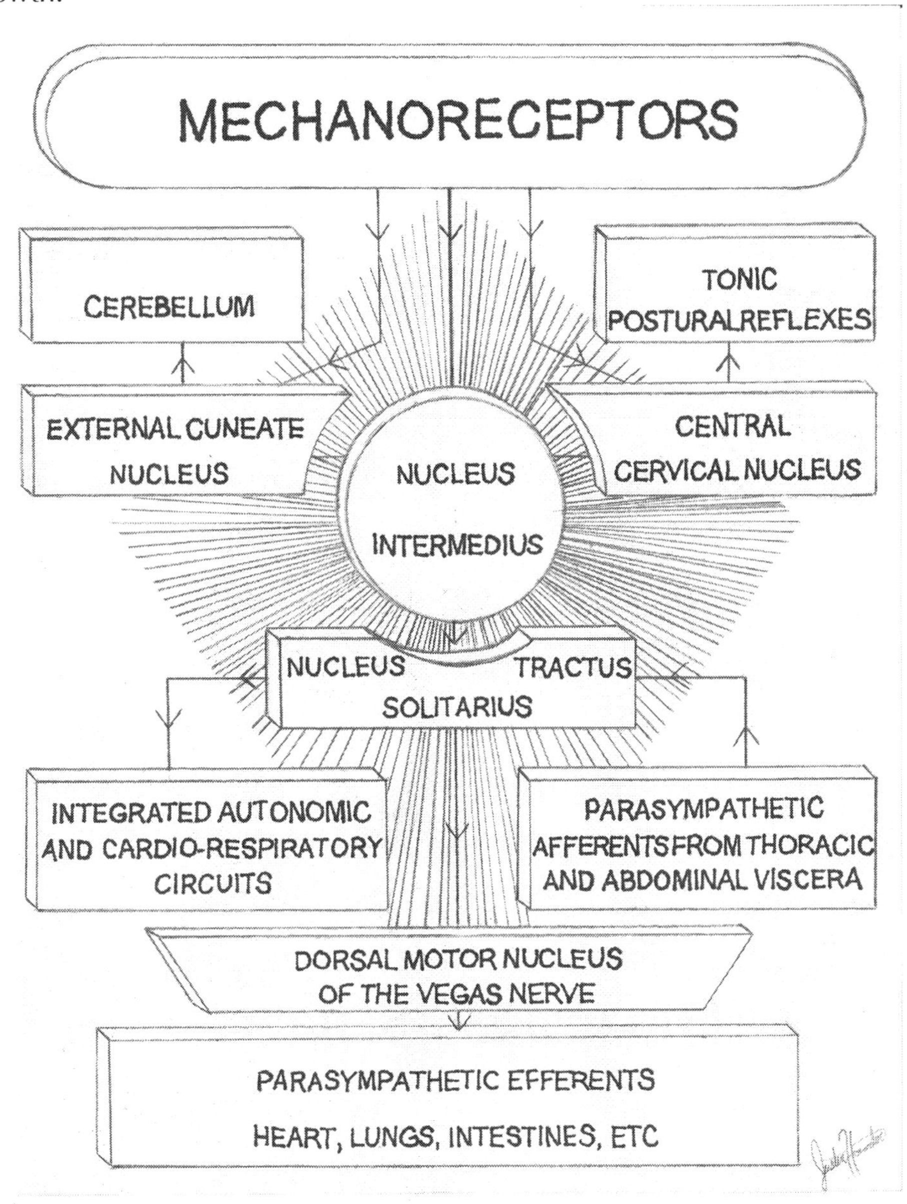

The Autonomic Nervous System

The Autonomic Nervous System is divided into two subsystems:
The Sympathetic Nervous System and the Parasympathetic Nervous System.
These two subsystems generally serve the same visceral organs; however, they cause essentially opposite effects. This counterbalancing of each system creates a balance of health. The sympathetic division mobilizes the body during experiences of fear (fight or flight). The parasympathetic division allows us to relax and conserve overall body energy (rest and digest).

The Sympathetic Nervous System

The sympathetic division of the nervous system, also known as thoracolumbar outflow, controls organs and glands during the fight or flight response. This is evident when we are in emergency or threatening situations or very excited. Signs of the sympathetic system at work are the heart pounding, rapid breathing, cold, sweaty skin, and dilated pupils. Less obvious are changes in brain wave patterns and in the electrical resistance of the skin. When sympathetic domination of the nervous system occurs, attention shifts from inside to outside. Blood flow is shunted away from deep organs of metabolism, digestion, and reproduction and shunts blood flow to muscles, eyes and the brain, which are mechanisms of survival.
The sympathetic division facilitates a number of adjustments to blood flow and glandular secretion during exercise. Skin (cutaneous) and visceral blood vessels (vessels that feed organs) are constricted, and vessels of the heart and skeletal muscles are dilated. Blood is caused to shunt to the skeletal muscles and to the vigorously working heart. The bronchioles in the lungs are dilated and glycogen and other stress hormones are mobilized from the liver and pancreas to increase blood glucose (sugar) levels. Nonessential activities of the body are placed in check. The sympathetic division's function is to provide optimal conditions for an appropriate response to a threat.
The stress response that shifts our resources so we can respond to challenge or threat is crucial for survival and often triggers healing processes in the body. The relaxation response is the polarity of this and reduces or ends the stress response as well as initiates constitutive healing processes in the body.
If this balance is upset by major stressors, relentless occurrences, or a lack of autonomic balance and responsiveness, we gain an increasing burden from stress. This burden is called allostatic load. This can have deleterious effects on many of our body systems that put us at risk, age us more quickly, and destabilize our moods and cognition.
Stress occurs when this sympathetic nervous system is stimulated consistently over time for inappropriate reasons, such as: traffic, basic human insecurity, fear, and worry. Fear comes in many flavors: fears of harm, fear to exist or not exist, fear of loss, abandonment or shame, fear of death, and many more varieties. It can arise in any chakra or area of life where we feel invested and at risk. Fear interferes with the process of logical thought and focuses the mind to the needs of survival. Fear is accompanied by blocks, especially in the chakras of the lower triangle, that inhibit wholeness, integration and spontaneous flow. Fear manifests physically as a contraction in the diaphragm, which alters breathing patterns.

The Parasympathetic Nervous System

The role of the parasympathetic division, also know as the craniosacral outflow, is to conserve the body's energy. It is sometimes called the resting and digesting system. When relaxing after

a meal, blood pressure and heart and respiratory rates are being regulated at low normal levels, the gastrointestinal tract is active in digestive activities, and the skin is warm. The eye pupils are constricted to protect the retinas from excessive light and the lenses of the eyes are accommodated for close vision.
This system is primarily concerned with healing and tissue repair. When stress levels are high and there is no relief from constant long-term pressure, normal repair and healing do not occur.

Structure

The roots of the sympathetic nervous system exit the spine from the first thoracic vertebra, located where your neck meets your head to the second lumbar vertebra, located just below the last ribs of your rib cage.
When there is dysfunction in the motion of this area of the spine the corresponding organ systems may be affected inappropriately. Shallow breathing habits and poor posture create abnormal pressure on the sympathetic nervous system. The classic examples are the asthmatic with a spinal dysfunction between the fourth and fifth chakra around the 3rd, 4th and 5th thoracic vertebrae, and the person with constipation or digestive maladies, which refer pain to the lower back. Referred pain occurs when a nerve is irritated and sends a powerful signal back to the brain and spinal cord. This is the somato (body)-sensory system. The typical pain in the left arm, jaw, and shoulder of a heart attack patient is an example of referred somatic (from the body) pain.
The roots for the parasympathetic nerves exit the spine in the upper neck and the base of the spine, the sacrum (sacred bone). Long deep breathing aids in accessing the parasympathetic nervous system.

Balance

The balance between the two nervous systems is important for all areas of human function. Kundalini Yoga tones the sympathetic nervous system and then allows the parasympathetic to heal during the relaxation phase of a yoga class.
Sex is one of the activities that require a balance of the parasympathetic and sympathetic nervous systems. Our very survival depends on this balance. Sexual arousal is only achieved in the parasympathetic nervous system, yet sexual completion is achieved with the sympathetic nervous system. People all develop their own ways to balance these two nervous systems during sexual contact.

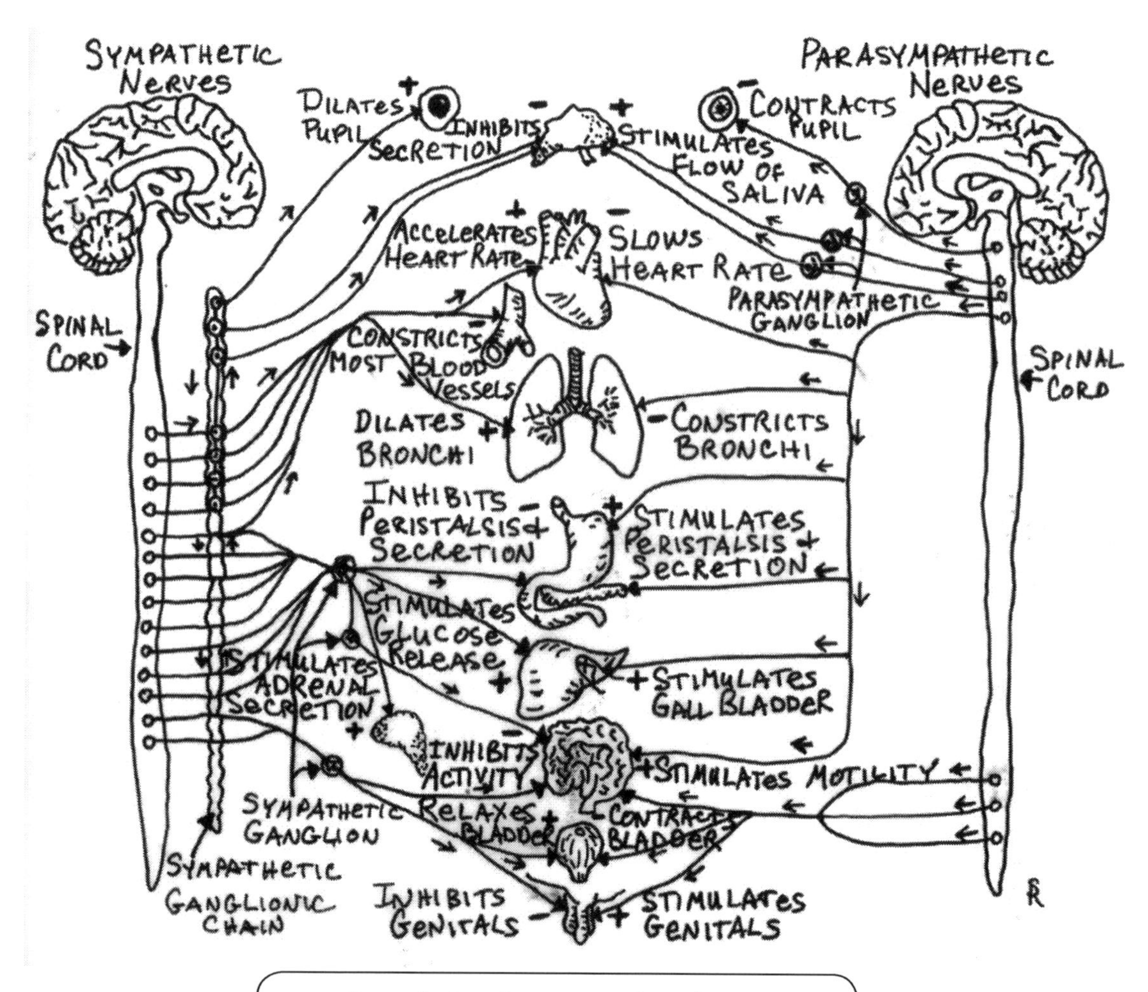

Sympathetic and parasympathetic division

Functions of the Sympathetic and Parasympathetic Nervous Systems		
Structure	**Sympathetic Function**	**Parasympathetic Function**
Eye	Dilates pupil (mydriasis); no significant effect on ciliary muscle	Contracts pupil (miosis); contracts ciliary muscle (accommodation)
Lachrymal gland	No significant effect	Stimulates secretion
Salivary glands	Inhibits secretion	Stimulates secretion
Sweat glands	Stimulates secretion (cholinergic fibers)	No significant effect
Heart:		
Rate	Increases	Decreases
Force of ventricular contraction	Increases	Decreases

Functions of the Sympathetic and Parasympathetic Nervous Systems		
Structure	**Sympathetic Function**	**Parasympathetic Function**
Blood vessels	Dilates cardiac & skeletal muscle vessels Constricts skin and digestive system blood vessels	No significant effect
Lungs	Dilates bronchial tubes	Constricts bronchial tubes Stimulates bronchial gland secretion
Gastrointestinal tract	Inhibits motility and secretion	Stimulates motility and secretion
GI sphincters	Contracts	Relaxes
Adrenal medulla	Stimulates secretion of adrenaline (cholinergic fibers)	No significant effect
Urinary bladder	Relaxes	Contracts
Sex organs	Stimulates orgasm	Stimulates sexual arousal

Homunculus and the Motor Cortex

The homunculus ("little man") is a traditional way of illustrating how the surface of the body is represented on the somatosensory cortex of the brain. Larger areas of the cortex are devoted to parts of the body that have greater sensitivity, such as the fingers, thumbs, and lips. The next time you are doing Kirtan Kriya, "Sa Ta Na Ma," remember the mental gymnastics involved.

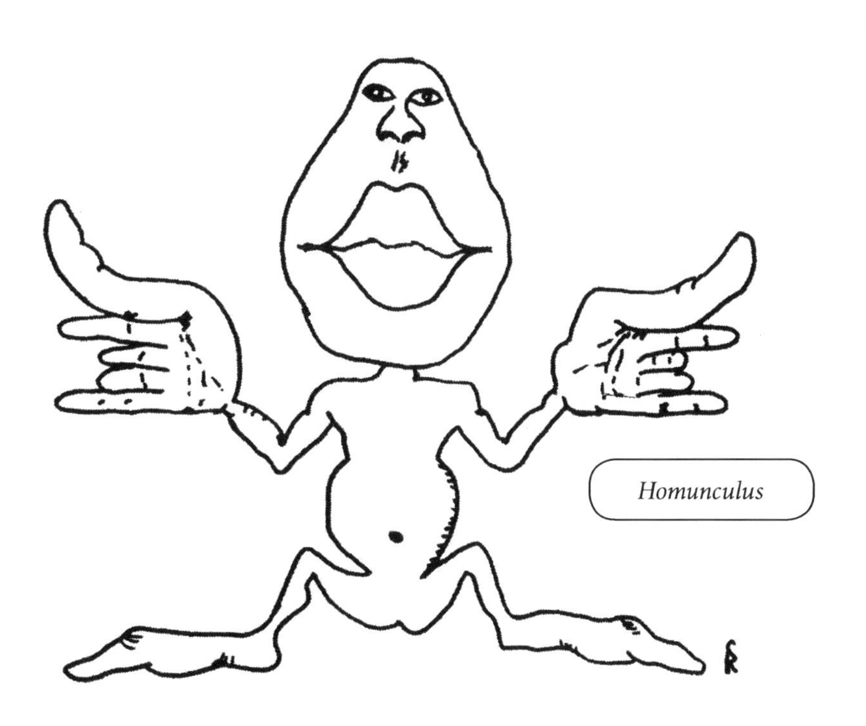

Homunculus

Three Major Developmental Parts of the Brain

The brain is an extremely complex organ with many interdependent parts, each having their own specific functions. However, these individual parts can be grouped into three developmental segments: the hindbrain, the midbrain and the forebrain.

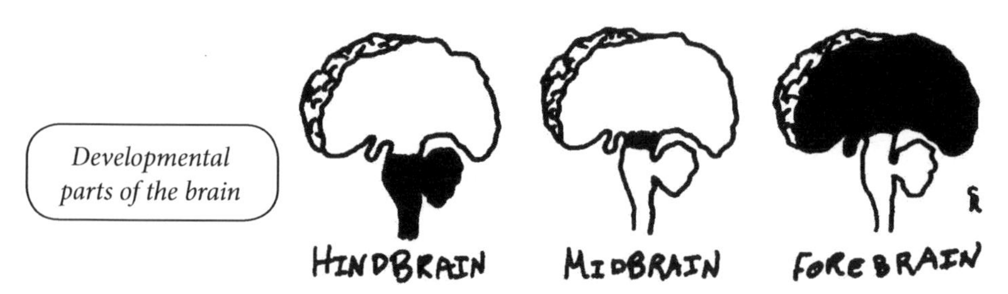

Hindbrain: It includes the upper part of the spinal cord, the brain stem, and the cerebellum. It controls the body's vital functions such as respiration and heart rate. The cerebellum coordinates movement and is also responsible for muscle memory.

Midbrain: It is the uppermost section of the brain stem. It controls some reflex actions, as well as certain voluntary movements. The midbrain, for example, is part of the reason why an eye is able to move.

Forebrain: It is the largest and most highly developed part of the human brain. The forebrain includes cerebrum, the hypothalamus, the thalamus, the basal ganglia and the hippocampus. It is from the forebrain that the "higher order" activities such as reasoning, remembering and thinking are derived.

Learning, Habits, and Patterns

Through the repetition of actions, the nervous system is educated and patterned. Many unconscious patterns are good and necessary. It is the interlock of a thought, and a subconscious feeling or belief which removes us from being present and real, that creates patterns of misperception, over- or under-reaction, and repeated patterns that increase rather than decrease suffering (i.e. more neurotic).

Neuroticism is also a temperament, which increases reaction to negative thoughts or obsess on a thought or feeling. The nervous system has the ability to compute on its own, but it will repeat a familiar pattern despite unpleasant results.

There are some patterns called reflexes that are hardwired into your spinal cord. For example, when you place your hand in a fire, the brain does not have to think about the color of the fire, the smell of the fire — it only removes your hand from the fire, faster than you know it happened. The sensory information from the burning hand goes to the spinal cord; the synapse occurs in the spinal cord and goes directly to the muscle removing the hand from danger. The brain is informed later. This is called a true reflex.

A Glimpse at the Relationship to Other Systems

All systems are directly or indirectly coordinated by the nervous system.

- Nervous system and glandular system work hand in hand to affect the chemistry of the brain. The hypothalamus acts as the intermediary for these two vital systems.
- Sends commands to muscles and organs.
- Monitors system function and regulates responses.
- Autonomic nervous system sends "go" or "relax" messages to all organs (lungs, intestines, heart, adrenals, etc.).

Negative Influences for This System

- Drugs, particularly cocaine, prevent the parasympathetic nervous system from doing its job.
- Marijuana affects the functioning of the brain.
- Toxins (like mercury) are common in our environment and cause damage to the nervous system

Good Habits for the Nervous System

Practice relaxing.

Meditation.

Exercise regularly

Use the nervous system in complex tasks.

Learn a foreign language

Listen to, or play music

Cold showers.

Yoga – Yoga stimulates nervous system to growth due to the demand put on the system.

Neural pathway development occurs in response to use and demand.

The actions of yoga create that demand to expand neural pathways.

Some Good Foods for This System

Trinity roots (onion, ginger, and garlic)

Herbal remedies: 3R, 4R, 26R, 18R

Celery juice

Phosphorus

Bananas

Bran tea

Calcium

Cucumber juice

Lecithin

Magnesium

Olives and good oils like flax seed oil, hemp oil & borage oil

Yogi Tea

Yogurt

Gotu Kola

Brain Health & Mental Agility Training

- Play music. Listen to music.
- Do math.
- Learn something new daily.
- Have new experiences regularly. Take a different route.
- Eat at a new restaurant.
- Listen to a new song.
- Meditate.

How Yoga Helps

Yogi Bhajan: "What is Kundalini? The energy of the glandular system combines with the nervous system to become more sensitive so that the totality of the brain perceives signals and interprets them, so that the effect of sequence of the cause becomes very clear. In other words, one becomes totally, wholesomely aware. That is why we call it the Yoga of Awareness. Kundalini is the creative potential of a person."

Yoga increases the energy moving through a particular nervous pathway (Breath of Fire particularly intensifies this).

Yoga balances the parasympathetic and sympathetic nervous systems.

Idea from yogic point of view is not necessarily to decrease the sympathetic stimulation, but increase parasympathetic ability to respond, to heal and adapt to change.

As we age we lose brain cells, gain others, and rearrange many networks. It is a complex process. There is active research being done on how many and what types. Since we only use a small percentage of our brain, perhaps those are the unutilized ones leaving more room for our functional cells. Perhaps this efficiency is a function of wisdom. It has been found that the interconnection between the brain cells, "neural plasticity," is what is important. This is strongly affected by exercise, robust circulation, good glandular response, and continuous activity and challenge.

Precautions Regarding Yoga and the Nervous System

This is one important factor in avoiding injuries in the practice of Kundalini Yoga! When the practitioner rises up quickly and drops his/her head back, the baroreceptor located in the neck (carotid) can be stimulated; the blood pressure in the brain is affected and often a loss of consciousness occurs. Do not wait until you hear the sound of thud!

If you are leading an exercise, alert fellow yogis to the possibility that they may feel a little lightheaded, and if they do they should sit or lie down with care.

To avoid any potential serious head or neck injuries, be especially careful in places with very hard floors such as concrete, tile, terrazzo, and hard woods.

Examples:

- Moving from Guru Pranam into Camel Pose (the classic example)
- Triangle to Cobra or Camel Pose
- Miracle Bend
- Stand, sit down, stand up, sit down, stand up, and inhale deep and lean back

Yoga That Affects the Nervous System

- Gong meditation, chanting, rhythm, sound are the best ways to access the parasympathetic nervous system
- Cold showers create a powerful parasympathetic reflex
- All exercises which involve rapidly changing hand and arm movements help to integrate the function of different areas of the brain, as well as moving the energy along the body's meridians
- Pranayam (Basic Breath Series, p. 81, Sadhana Guidelines)
- Breath of Fire helps "tone" the sympathetic nervous system
- Left nostril breathing activates calming, parasympathetic simulation and the intuitive,

creative faculties of the right hemisphere of the brain

- Right nostril breathing activates the stimulating sympathetic system and the action-oriented, analytical faculties of the left hemisphere
- Alternate nostril breathing balances the activity of the two branches, giving an experience of altered receptivity and readiness for action, and integrating ideas into action
- To Activate the Central Nervous System and Stimulate the Pituitary Gland Kriya (p. 15, Physical Wisdom)
- Nabhi Kriya
- Sat Kriya
- Kirtan Kriya
- Sarabandandi Kriya
- Stretch Pose
- Triangle Pose
- Kriya to Balance the Parasympathetic and Sympathetic`

The Essential Nervous System in Kundalini Yoga —

The Sensory Human: Maintain your awareness! Perceptions are experiences we obtain through the senses: smell, taste, sight, touch and hearing. The senses are directly related to the sensations we experience, and as Yogis this is the pallet of living. Our brain has a similar experience, whether the experience is real or only imagined in the mind. Every thought, emotion and image creates molecules in the brain that affect the glands. Buddhist monks have learned to temper outside experience with inside mantra, thus being less reactive to the environment and more stable. This requires a strong nervous system.

What is a strong nervous system? There are several ways to measure the strength of the nervous system. The origin of the message is the mental impulse, guided by the psyche. The ability to hold a thought over time is an acquired skill. Dropping a thought as well is a skill necessary for the brain.

Shaking all over? The mental signal travels down the cervical spine and the spinal cord. Most nerves synapse at this point as they exit the spine. This is the beginning of the peripheral nervous system and these nerves have insulation on the outside. If this sheath is not adequate, the signal is interrupted and shaking of a muscle or limb may occur. Shaking when you are stable occurs when the practitioner drives a high voltage signal in a consciously commanding manner, resulting in a shaking of the surrounding muscles. This is one of the oldest methods of raising the Kundalini.

Stability. The ability to respond like a thunderbolt is another form of strength. The ability to be stable and not react is a form of strength.

The ability to hold a thought over time. The mind can hold a theme that can prevail over time. The mantra Har har har gobinday, har har har mukhunday ... aids in this pursuit.

The stability and sensitivity of an Aquarian human: ratio and proportion. Kundalini Yoga and Meditation helps develop a strong nervous system. A strong nervous system sends and receives signals, from brain cells to tissue cells and back to brain cells. Being a responsible person is about being able to respond, a function of command, royal courage, and stability. The more mental and nervous system stability one has, the more one becomes a sensitive being, also known as the sensory human.

Adaptability, the key to evolution and success. Any change requires the ability to drop the past and create new neurological pathways. The ability of the brain to change and adapt to change is called neuroplasticity. When we fire neurons as a group, they create and permeate pathways. This is the path of self-discipline, the sadhu and the yogi.
Special skills of brain cell management, neuro-patterning and the electromagnetic field of attraction. The movements we use in Kundalini Yoga have the effect of rewiring the brain by utilizing pathways repetitively. A good example is the way we use the breath and the navel point. Usually we move in parallel motion, and sometimes we separate the two motions. This is a big step in being able to speak and chant effectively while breathing. Within the psyche, Kundalini Yoga separates the life force energy of the breath and the neurons that connect the diaphragm to the brain and the navel point, the place your original nourishment comes from, the center of your pranic nervous system, where all 72,000 nadis come together. Separating these regions of the mind is a step towards enlightenment. Doei Shabd Kriya is a great meditation that facilitates brain development. Breath walk is also great for brain balancing and the dance of the homunculus.
Divine input / sublime output. Your brain operates all the time, even more actively when we are sleeping. It remembers everything! During your conscious, waking life, everything is recorded. During your time of sleep you are still aware, listening to everything. The environments you expose yourself to during your aware time are permanently recorded. During sleep, Yogi Bhajan suggested listening to audio mantra, so your mind can develop an anchor.

Some Conditions of the Nervous System

Stroke: Leaking blood into the brain.
Meningitis: An infection of the spinal fluid.
Multiple Sclerosis: The body attacks the protective sheath that protects the nerves.
Alzheimer's Disease: A disease that becomes apparent when memory loss occurs.
Parkinson's Disease: A lack of dopamine in the brain.

Vocabulary

Afferent: The nerves that send sensory information from the peripheral nervous
Baroreceptor: A receptor that is stimulated by pressure changes.
Chemoreceptors: Two distinct classes of chemoreceptors are recognized: taste (gustatory) receptors, i.e., the taste buds of the tongue and smell (olfactory) receptors, embedded in the lining of the nasal cavity.
Efferent: The nerves that send motor information from the central nervous system(the brain) to the peripheral nervous system.
Ganglion: The collection of nerve cell bodies outside the CNS. Most ganglions are associated with chakras.
Hypothalamus: The part of the diencephalon that regulates many basic body functions such as temperature regulation. It is the gray matter in the floor and walls of the third ventricle of the brain.

Mechanoreceptors: A type of sensory reception that is subdivided into sensitivity to touch, pain, sound, gravity, muscle tone, and vibration. It is the basis for reflexes, such as the patellar reflex(knee).

Nociceptors: Receptors or free nerve endings associated with pain.

Paraganglionic neuron: The autonomic motor neuron that has its cell body in the CNS and projects its axon to a peripheral ganglion.

Photoreceptors: The cells that absorb light energy and transform them into electrical impulses allowing for the sense of sight.

Postganglionic neuron: Automatic motor neuron that has its cell body in a peripheral ganglion and projects its axon to an effector cell or organ.

Receptor (or sense cells): A specialized structure of the sensory nerve that is excited by a certain stimulus.

Sensory reception: The mechanism by which a person is able to react to external or internal environmental changes.

Sympathetic tone: The state of partial vasoconstriction of the blood vessels maintained by sympathetic fibers.

Thermoreceptors: The sensors that make us sensitive to external and internal environmental changes. They stimulate behavioral responses such as sweating and shivering. They help us maintain a constant body temperature. These receptors are found in the skin, in deep tissues of the body, and in the hypothalamus and the spinal cord (hypothalamic thermostat).

Study Points

1. What is the effect of the sympathetic nervous system on the heart, lungs and eyes?
2. How does yoga strengthen the nervous system?
3. How does stress effect us and why?

Chapter Three

The Spine

The shape of the spine is a reflection of a person's psyche, breath pattern, and habits. These factors express themselves as one's attitude, posture, and projection.

Basic Structure

Spinal Column

The spinal column is composed of vertebra stacked on top of each other with a disc in between.

It is divided into 3 specific sections:

C - Cervical, B - Thoracic, A - Lumbar

CERVICAL NECK — C

THORACIC & ARTICULATE WITH THE RIB CAGE — B

LUMBAR REGION — A

Spinal column

Vertebra and Disc

Vertebra

Each vertebra is made of a body that is shaped like a barrel and has two additional joints in the posterior for stability.

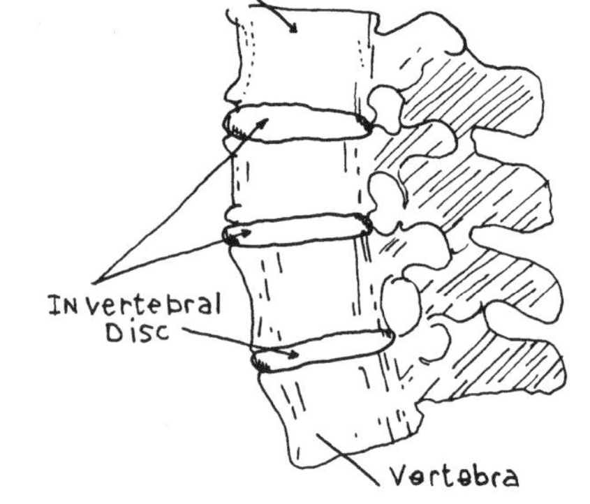

Disc

The cushioning and connecting tissue between the vertebras is called the disc. The disc consists of a fibrous outer wrapping called the annulus, and an inner core called the nucleus.
When the outer annulus bulges or rips, the nerve roots or spinal cord can be crushed, causing great pain. The discs degenerate from excessive load, smoking, and poor circulation. The disc tissue holds water and uses this ability (hydrostatic pressure) to remain resilient and supple. When discs hold water they function properly.

When they lose water they become brittle and more prone to injury. Spinal twists, as well as spinal flexion, are good exercises to maintain good spinal health.
The disc cushions the space between the vertebral bodies and a synovial capsule surrounds the posterior joints. There are many muscles attached to the spine to provide movement in all possible directions. Some muscles are only an inch or so long and some extend the entire length of the spine. The first two vertebras below the skull are unique. They have no stabilizing facets and are structurally vulnerable due to their inherent instability.

THREE SECTIONS OF THE SPINAL COLUMN

1. Cervical (neck):

The top seven vertebrae support the weight of the skull (11-12 lbs.) and provide mobility to the head.

The Upper Cervical Spine

Atlas and Axis: The first (top) vertebra is called the atlas (C1) and the second (C2) is called the axis. Both C1 and C2 (atlas and axis) have unique structures that allow free motion in the upper neck and provide the stability required to keep the head on straight.

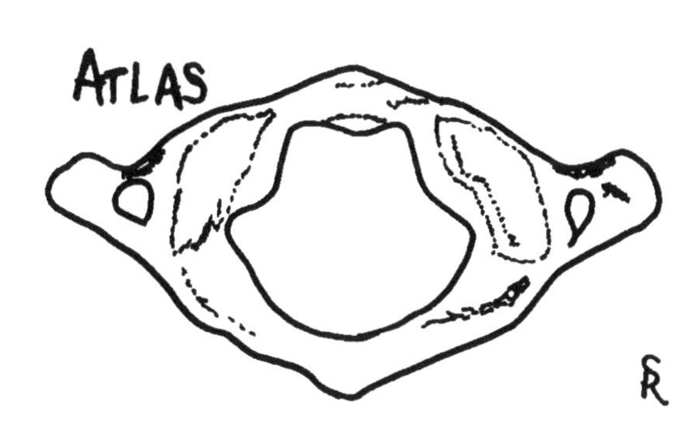

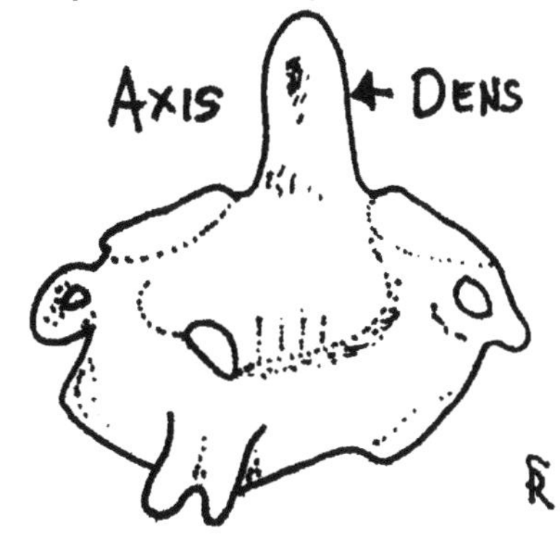

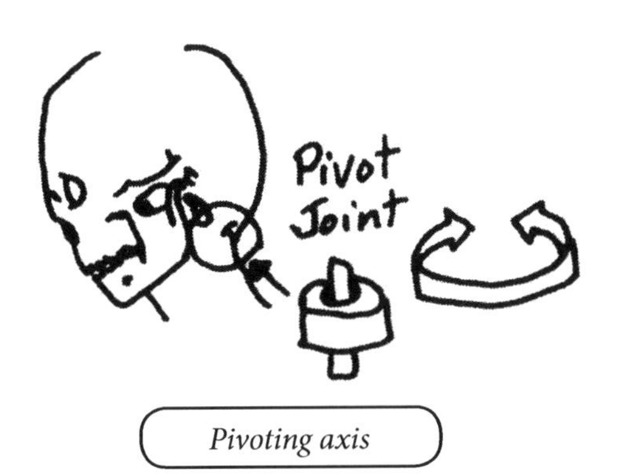

Pivoting axis

There is no fibrous disc between the skull and C1, OR between C1 and C2. This allows for great freedom of movement but little structural stability.
The atlas and axis are the most vulnerable to injury and have the highest concentration of mechanoreceptors. The nerves are largest as they pass through the brainstem and the passageway is the most restricted in this region.

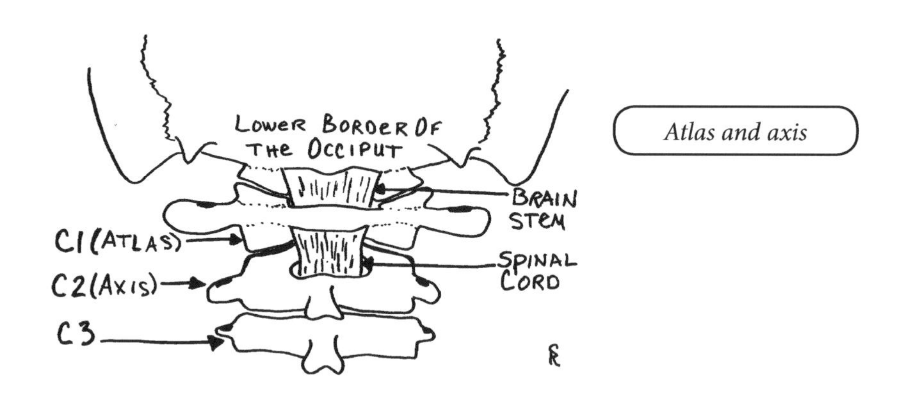

Atlas and axis

The atlas acts mechanically like a washer and the axis acts like a pivoting axis for rotation. The transverse and alar ligaments help keep C1 in place as it moves around the dens of C2.

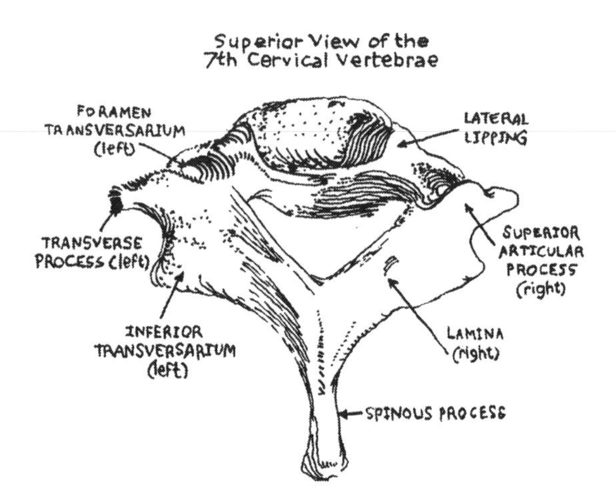

1. The Lower Cervical Spine

The lower cervical vertebrae all have a body, disc, transverse process, and two facet joints. The cervical spine is relatively mobile providing much of the nodding motion.

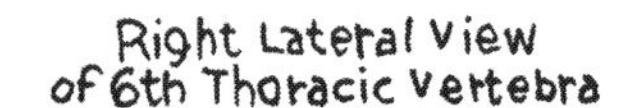

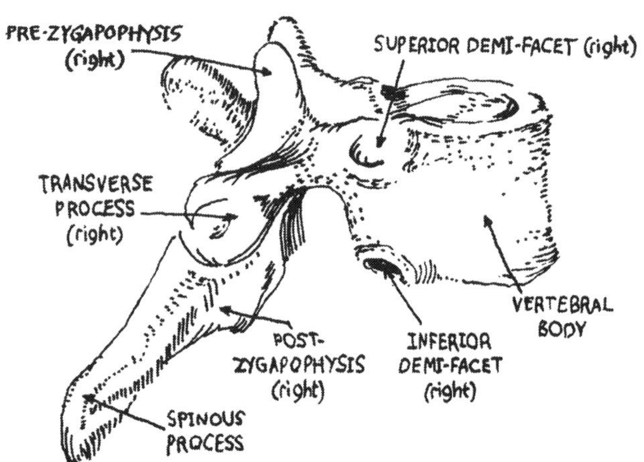

2. Thoracic:

The middle twelve vertebrae articulate with the rib cage. The shape of the thoracic region determines the shape and mobility of the rib cage facilitating the breath.

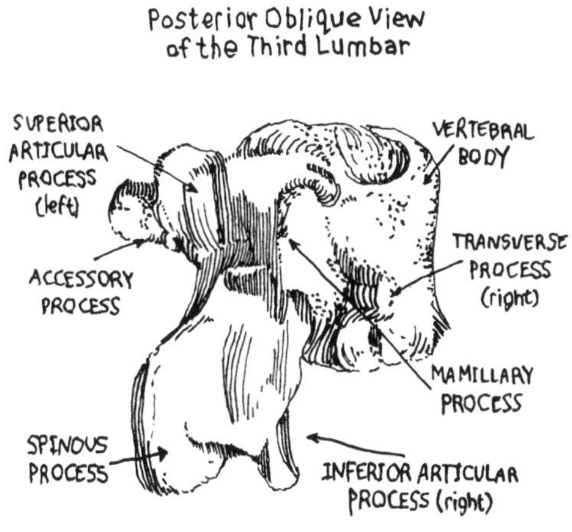

3. Lumbar:

The low back area is the lumbar region containing five vertebrae.

Fourth lumbar: The fourth lumbar is often referred to as the seat of Kundalini (counting up from the base). This is what anatomist calls the second lumbar vertebra(counting down from the top lumbar).

Pelvis

The pelvis includes the sacrum, 2 hip bones, and the coccyx.
Sacrum: These 5 fused vertebrae are the base of the spine. The sacrum normally fuses between the ages of 16-18.
Hip bones: Two bones made up of 3 sections, the ilium, ischium and pubis.
Coccyx: The coccyx is at the base of the sacrum. It is also know as the tailbone and consists of 3-5 fused vertebrae.

Skull

At the top of the spine sits the skull (cranium) that, together with the facial bones, comprises the skull. The cranium has 26 bones (some books say from between 22 to 29 bones depending on how you count them). The bones of the cranium have specific types of joints that allow the cranial cavity to move slightly. This motion is referred to as cranial respiration.

7 + 12 + 5+ Pelvis & Skull = 26 motion segments in the spine

Scoliosis of the Spine

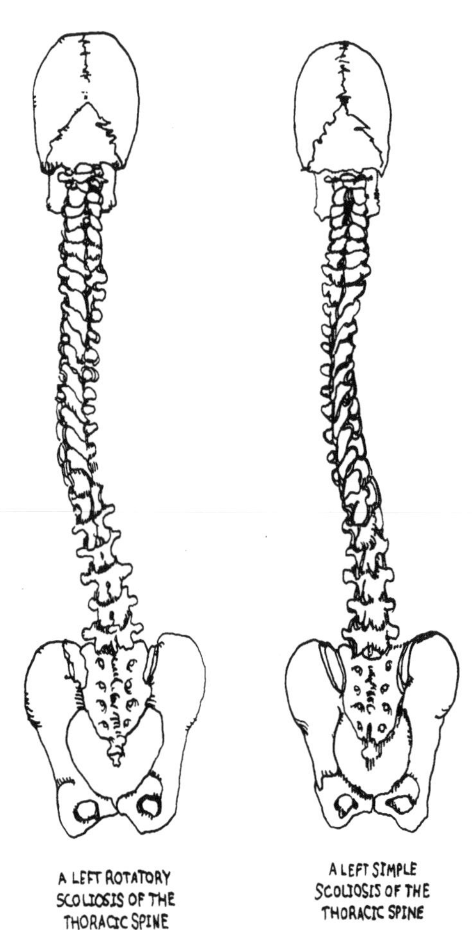

A LEFT ROTATORY SCOLIOSIS OF THE THORACIC SPINE

A LEFT SIMPLE SCOLIOSIS OF THE THORACIC SPINE

Spinal Curves

At birth, the spine has a curve with the concave portion facing forward like a C shape. This is known as the primary curve. It forms the curve in the thorax(and ribs) and sacrum.
The first secondary curve occurs in infancy, at the neck when the head is raised to seek food.
The second secondary curve develops in the lower back and forms during the walking stage.
The dynamics between the primary and secondary curves are intricate to the engineering of this weight-bearing structure.
When there is a curve laterally, to the side, this compromises the primary and secondary curves and creates a condition known as lateral spinal curvature, or scoliosis.

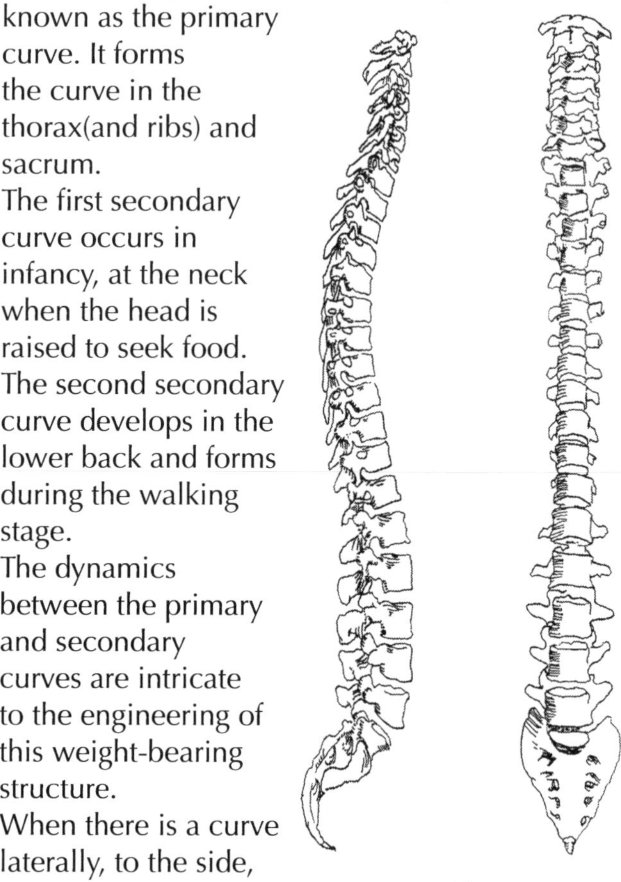

Normal Spine

Basic Function

The spine has two mutually exclusive functions that it needs to balance. The first is housing and protecting the nervous system; the second is allowing for mobility. When there is a choice, protecting the nervous system wins, and the muscles connected to the spine go into spasm to prevent mobility.
The human spinal (vertebral) column is a highly versatile mechanism that displays all the characteristics required of a crane, such as rigidity, strength and leverage. It is, however, extremely elastic and flexible. The spinal column exhibits more varied functions than any other anatomical unit of the human body.

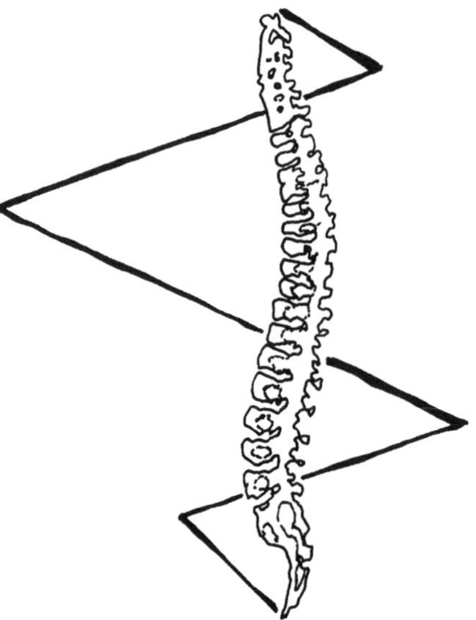

LATERAL VIEWS OF THE SPINAL COLUMN INDICATING THE FOUR SPINAL CURVES

Stabilization

The vertebrae and their attached muscles play an important role in the stabilization of the entire body.

Support and Weight Bearing

The spinal column is the support of the head and the upper extremities. It also bears the weight of the entire upper body. The sacrum and pelvis transfer the force of gravity to the legs.

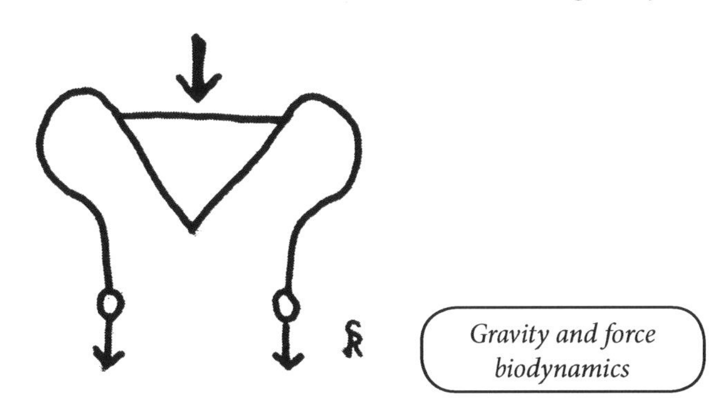

Shape and Position

The spinal column gives proper body contours and erect posture.

Motion

Allowing flexible movement is an important function of the spinal column. Spinal movements are flexion, extension, lateral flexion, and rotation.

Skeletal Formation

The spinal column is part of the axial skeleton. The thoracic region of the spinal column forms a portion of the posterior thoracic wall. This also provides attachment for the ribs.

Resiliency

The ability to absorb shock is a very important function of the spinal column. Otherwise, damage might occur to the visceral structures it protects, the brain and the spinal cord in particular.

Why Spinal Flexibility Is So Important!

Just as nerve roots exit the spine to send information from the brain to other tissues (glands, organs, etc.) there is also a reciprocal input system. This system takes information from organs, glands, and motor units and sends the messages back to the brain.

Between each vertebra is a series of important structures, joints, ligaments, and tendons that all send positional information to the spine. The areas of the spine that move less have less healthy neurological messaging and possible impaired function. When the joints in the spine move in a pathologic (altered motion) manner, this sends a pathologic (sick) message to the brain. Poor input = poor output! This message affects the nerve roots in that region as well as the nervous system as a whole and can create a condition known as somatic (whole body) dysfunction.

How Yoga Helps

- Spinal flexibility facilitates a full, deep breath.
- There is a direct relationship between breathing and spinal flexibility. A full breath will move the spine in harmony and creates a resonance that circulates the spinal fluid with a strong and gentle wave motion.
- Flexibility of the spine will help the joints of the spine stay healthy. Yoga helps the disc maintain the right amount of hydration, flexibility, and strength.
- The spine is rich in mechanoreceptors. These specialized nerves send impulses to the brain and aid in perpetuating normal homeostasis.

Precautions Regarding Yoga and the Spine

- Most injuries occur while bending and twisting in combination.
- Always do neck rolls slowly. Move in and out of Plow Pose slowly and with control.
- Move slowly into poses of spinal extension such as Cobra Pose, Wheel Pose, and Bow Pose.
- Special care must be taken during movements of the upper cervical spine. The atlas (C1) has a great range of motion in all directions because it does not have the connecting joints and discs that the rest of the lower neck does. The atlas pivots freely around the axis (C2), making it very important for motion of the head and also more vulnerable to injury or misalignment.
- Headstands that put force on the top of the head are not recommended for several reasons.

Yoga That Affects the Spine

- Basic Spinal Energy Series (p. 45-46, Sadhana Guidelines)
- Lower Spine & Elimination Set (p. 52, Sadhana Guidelines)
- Flexibility and the Spine (p. 47-50, Sadhana Guidelines)
- Exercises that strengthen the abdominal muscles
- Spinal Flex
- Cat-Cow
- Cobra Pose

Some Conditions and Injuries of the Spine

Subluxation: Two vertebral segments malpositioned, which results in an interruption of the nerve flow. This may affect both the afferent and efferent nerves resulting in illness, malfunctions and pathology.
Simple sprain: Strain of the muscles surrounding and supporting the spine. This is the most common injury and it is usually innocuous. It can be treated with ice and rest. If pain persists or is severe, consult with a health care professional.
Pinched nerve: Usually causes a burning sensation. A pinched nerve may refer pain. An injured disc often causes this type of pain. Severe spinal pain with radiation down one or both of the legs may be a medical emergency.
Cancer: Spinal pain from cancer should be dealt with great care, using slow breath meditations and deep relaxation techniques.
Laminectomy: The surgical removal of the vertebral lamina, most often done to address the

symptoms of a ruptured disc.
Lumbago: Aute lower back pain.
Spina Bifida: A congenital defect resulting from the failure of the vertebral bones to fuse correctly.
Spinal Fusion: The surgical removal of the disc and insertion of bone strips which grow together to make two vertebrae into one.
Ankylosing Spondylitis: A condition where the vertebrae of the spine fuse into one bone, just as the sacrum did during youth.
Accentuation of Normal Spinal Curves:

- Excess - Lordosis of the cervical spine or lumbar spine
- Excess - Kyphosis of the thoracic spine
- Reverse Curves in the Spine:
- Military Cervical Spine: Loss of normal curve.

Spondylollsthesis: Either anterior or posterior vertebral displacement, they range from mild to severe.
Scoliosis: A lateral curve of the spine. Most often found in teenage girls. It can range from mild to severe.

Vocabulary

Arachnoid: (Anatomy) designating the middle of three membranes covering the brain and the spinal cord.

Cervical: Of the neck

Choroid plexus: A capillary knot that protrudes into a brain ventricle; involved in forming cerebrospinal fluid.

Disc: A layer of fibrous connective tissue with small masses of cartilage among the fibers, occurring between adjacent vertebrae.

Dura mater: The outermost, toughest, and most fibrous of the three membranes covering the brain and spinal cord.

Lumbar: The part of the back just below the thoracic region.

Pia mater: The vascular membrane immediately enveloping the brain and spinal cord and surrounded by the arachnoid and dura mater.

Process: A projection or outgrowth from a larger structure, usually a bone (the alveolar process of the jaw).

Thoracic: The area of the trunk covering the part of the body cavity from the neck or head to the abdomen, containing the heart, lungs, etc.

Spinal column: The series of joined vertebrae forming the axial support for the skeleton.

Spinal cord: The thick cord of nerve tissue of the central nervous system, extending down the spinal canal from the medulla oblongata.

Synovial capsule: The clear, albuminous lubricating fluid secreted by the membranes of joint cavities; lubricates joint surfaces and nourishes articular cartilage.

Study Points

1. How many vertebrae are in the spine?
2. What are the 3 regions of the spinal column?
3. What area of the spine has the most mechanorectors?
4. How does spinal flex effect the discs of the spine?

Chapter Four

Cardiovascular System

"The heart beat begins the song of life."

Basic Structure/Function

The Cardiovascular System is also known as the Circulatory System. The Cardiovascular System consists of the heart (cardio) and the blood veins (vascular) that transport nutrients and oxygen through the body, and remove toxins and waste away from the body's organs.

The heart pumps the blood through the body transporting oxygen, carbon dioxide, hydrogen, heat, antibodies, clotting factors, hormones, water, nutrients and wastes, to and from the body's cells. The cardiovascular system is the "highway of health" which delivers the fuel and nutrients to the cells of the body. It is a highway of chemicals. It plays an integral part in regulating fluid volumes of the body, maintaining the acid-base balance of the system, regulating body temperature, hormonal communication, and protecting against infection.

Organs in This System

Heart, blood, arteries, veins, capillaries

-OR-

The pump, the liquid, and the plumbing

Heart: The Double Pump

A muscular, pear-shaped organ, the heart is the center of the circulatory system.

The heart consists of several layers of a tough muscular wall called the myocardium.

A thin layer of fibrous membrane, the pericardium, covers the outside, and another layer, the endocardium, lines the inside.

The heart is divided down the middle by a wall of muscle into a right and left side, which in turn are subdivided into upper and lower chambers.

The upper chamber is called the auricle (also called the atrium) and the lower chamber is called the ventricle.

The two auricles act as receiving chambers for blood entering the heart; the lower, more muscular ventricles pump the blood out of the heart to the body and the lungs.

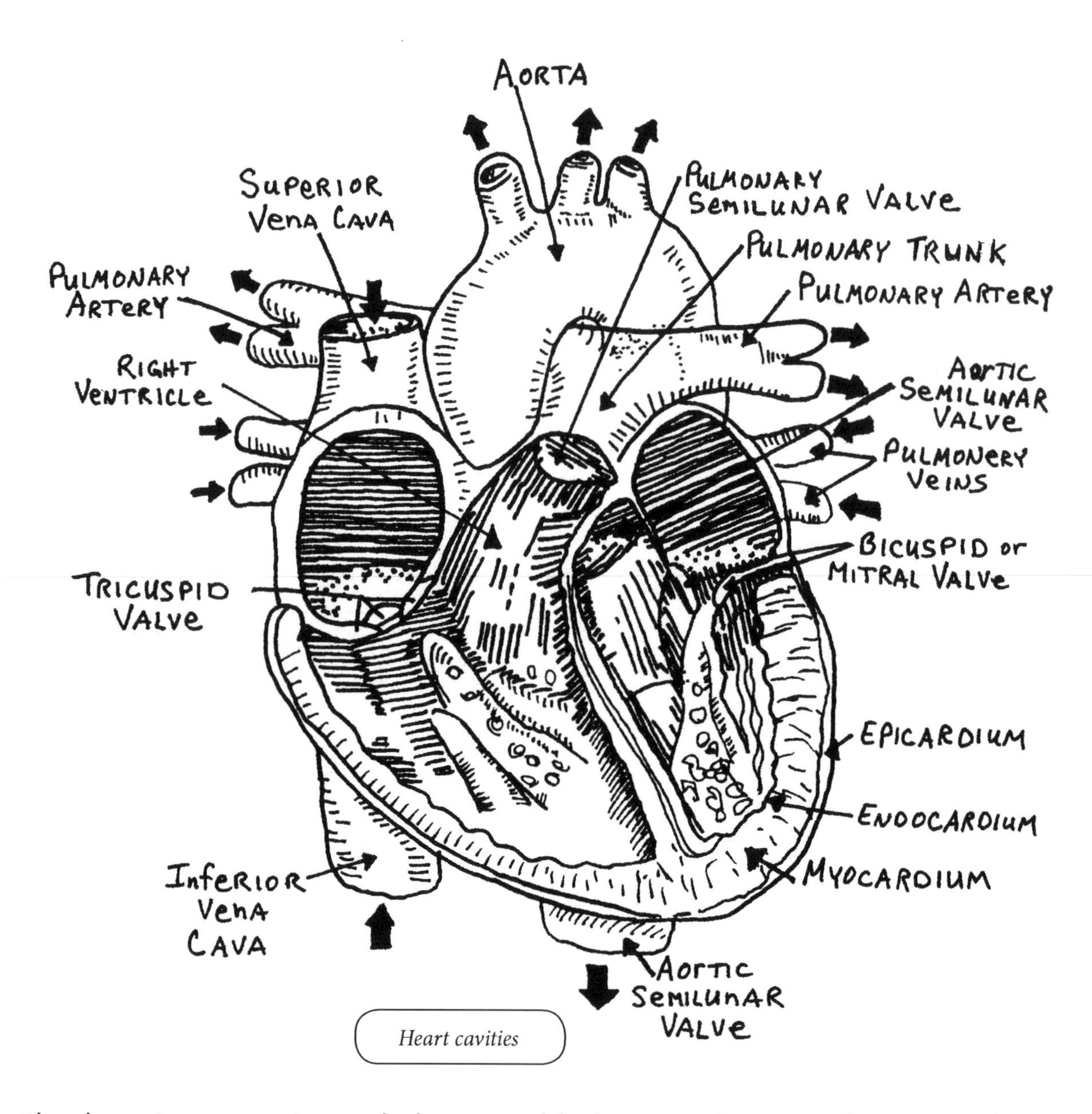

Heart cavities

The alternating contractions and relaxations of the heart muscle (myocardium) pump the blood to the body and lungs and create the heartbeat. These contractions are initiated by electrical impulses from a natural, built-in, automatic pacemaker, the sinoatrial, or S-A node located in the muscle of the right auricle.

An impulse from the S-A node causes the two auricles to contract, forcing blood into the ventricles.

Contraction of the ventricles is in turn controlled by impulses from the atrioventricular, or A-V, node, located at the junction of the two auricles.

Following contraction, the ventricles relax and as pressure within them falls, blood again flows into the auricles.

An impulse from the S-A node starts the cycle over again. This process is the cardiac cycle.

The Flow of Blood Through the Heart

The left side of the heart pumps blood to the entire body, and the right side pumps blood to the lungs.

THERE ARE FOUR HEART VALVES.

Deoxygenated blood that has circulated throughout the body enters the heart through the right auricle (1).

The blood then passes directly through the tricuspid valve into the right ventricle (2) and is then pumped through the pulmonary artery to the lungs for re-oxygenation.

Simultaneously, fresh, bright, electric, fiery red, oxygenated blood from the lungs enters the heart through the left auricle (3) and flows through the bicuspid (mitral) valve into the strong muscular left ventricle, and is then pumped into the aorta (4) and out into the arteries of the body.

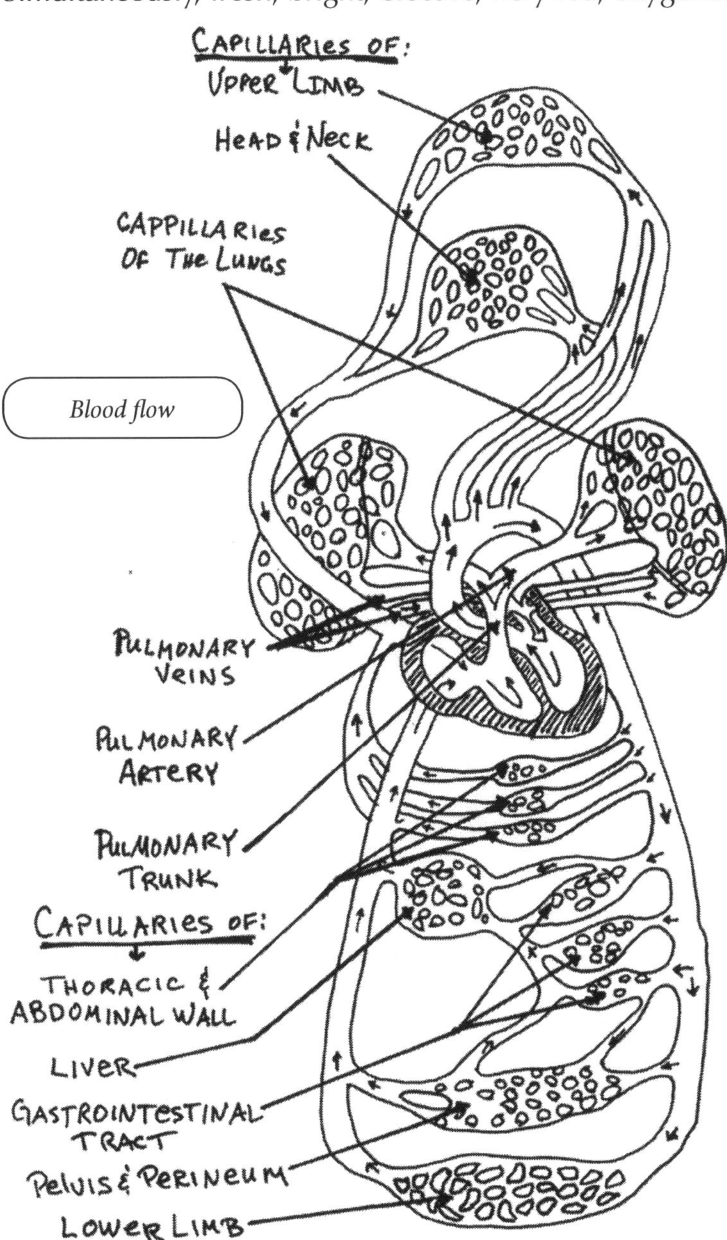

The muscular wall of the left ventricle is the thickest area of the heart muscle because it has the biggest job: pumping blood throughout the entire body. Valves in the heart allow blood to flow in one direction only and, in addition, help maintain the tremendous pressure required for blood to be pumped throughout the body.

The heart is innervated by the vagus nerve (cranial nerve X), one of the twelve cranial nerves. The vagus nerve has a parasympathetic (relaxing) effect on the heart and controls heart rate. The heart is also innervated by sympathetic (fight or flight) nerve fibers, which serve to increase the heart rate. These two types of nerve fibers regulate the conduction system, and speed of the heartbeat.

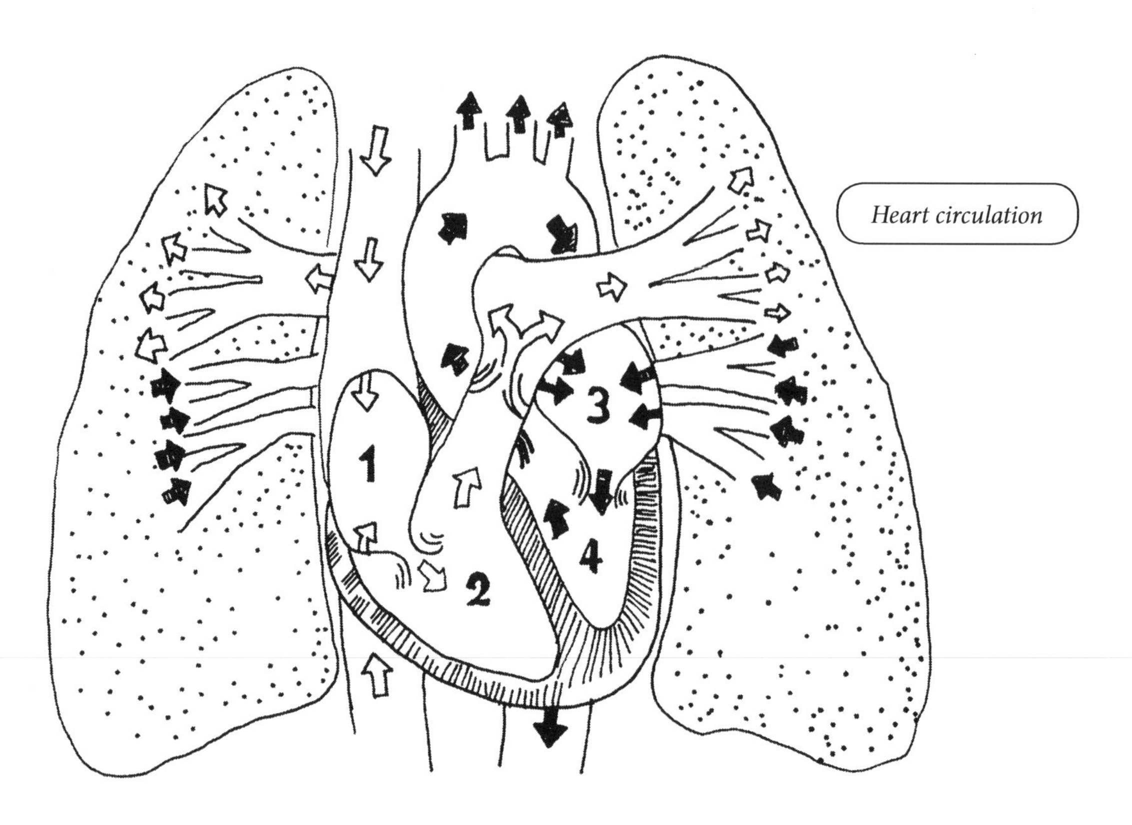

Arteries

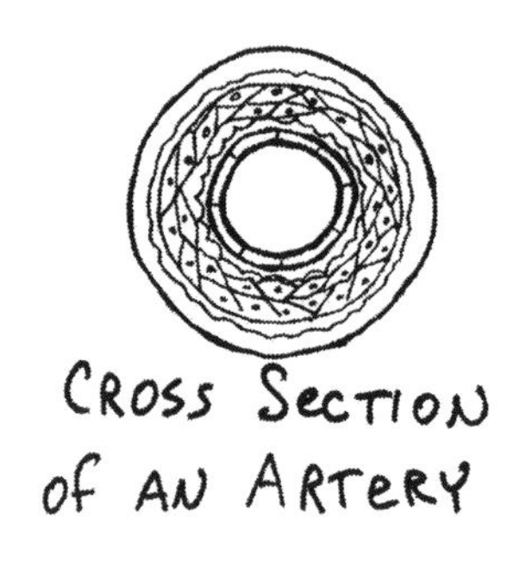

The arteries transport blood away from the heart. The arteries are more organized than the veins, and the structure of the walls is thicker, to handle the higher pressure. The smooth muscle layer acts to regulate the distribution of blood.

Foods that adversely affect the arteries, causing them to clog, are meats and saturated fats. Smoking tobacco (very high in its concentration of toxic chemicals) creates deficiencies of vitamin C, dehydration, increases stress, and can cause chronic high blood pressure, all of which have a negative effect on the arteries.

Veins

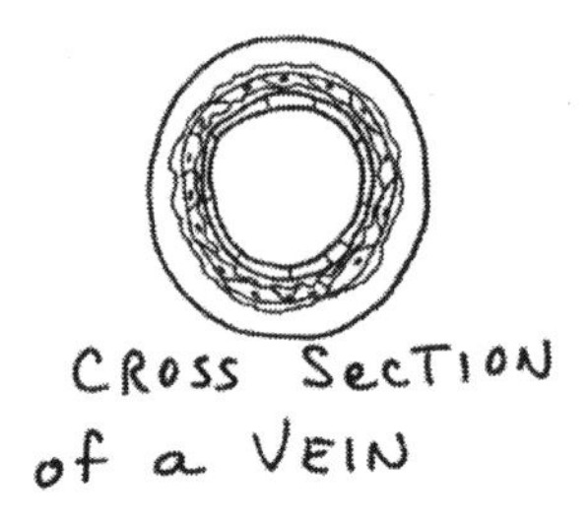

The veins deliver blood under lower pressure than arteries and return it to the heart. The veins have less muscle and elastic tissue in their walls, allowing them to stretch to become a virtual reservoir of blood. The simple tissue of the veins (endothelium) forms valves at certain points in the venous system of the limbs and neck, which prevent backward flow and resist blood pooling in the lower extremities.

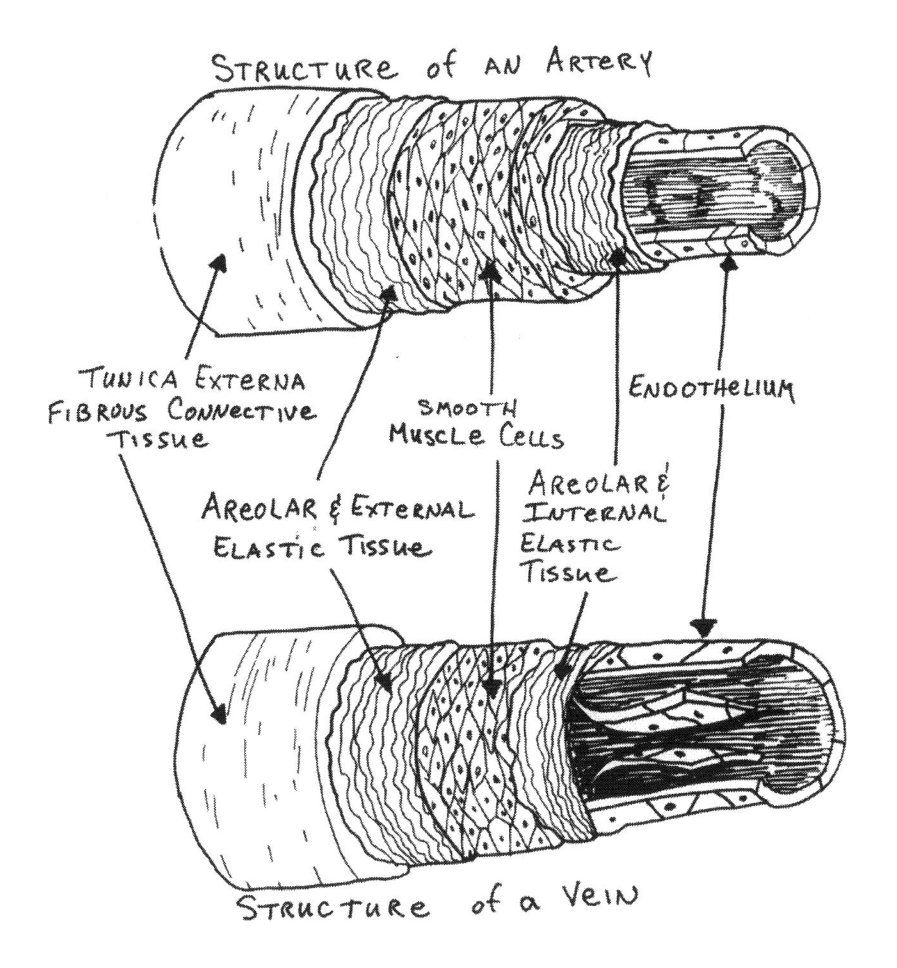

Capillaries

The capillaries are simple endothelial tissue tubes. The nutrients and gases can diffuse into the tissues by simple diffusion and osmotic pressures. The dilation and constriction of the capillaries is mediated by the nervous system as an adaptation to the environment and the needs of surrounding tissues.

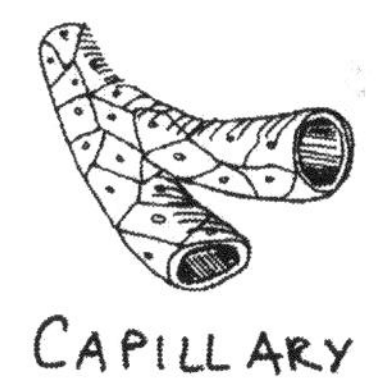

Blood

The blood is composed of 45% red blood cells and 55% plasma. Red blood cells transport oxygen through a molecule called hemoglobin. The blood contains approximately 4.5 million erythrocytes, red blood cells, which are produced in the marrow of the bones. The blood also carries leukocytes, white blood cells, which help produce the antibodies to fight infection. The third group of blood cells is platelets.

ERYTHROCYTES

Erythrocytes (Red Blood Cells)

Erythrocytes, or red blood cells, are formed in the bone marrow. When mature, they are literally sacs of hemoglobin, the iron-containing pigment that transports oxygen in the blood.

Leukocytes (White Blood Cells)

Leukocytes, or white blood cells, defend the body against infections and foreign materials. The number of leukocytes in the blood is often an indicator of disease. There are several types of leukocytes. Granular leukocytes include neutrophils, basophils, and eosinophils. Nongranular leukocytes include lymphocytes, monocytes, and macrophages.

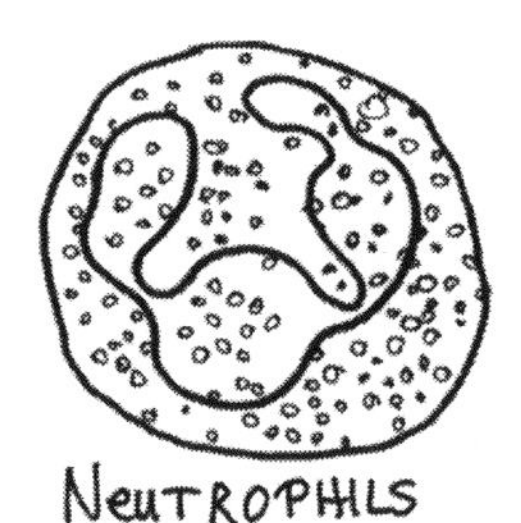

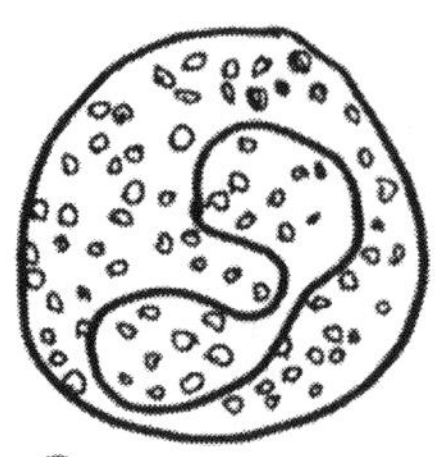

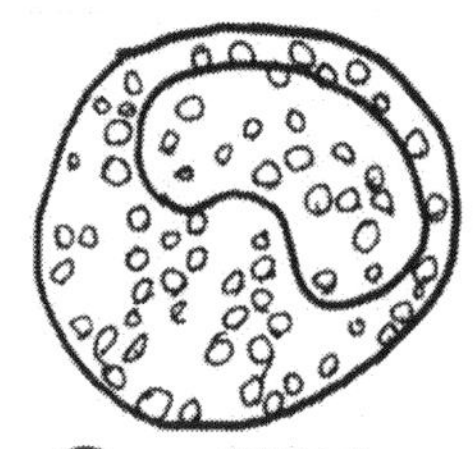

Neutrophils defend the body against bacterial or fungal infection. Pus is the result of neutrophil activity.

Eosinophils primarily defend against parasitic infections. They are the predominant inflammatory cells in allergic reactions such as asthma, hay fever, and hives.

Basophils release histamines, causing inflammation, as an allergic or antigen response.

Lymphocytes

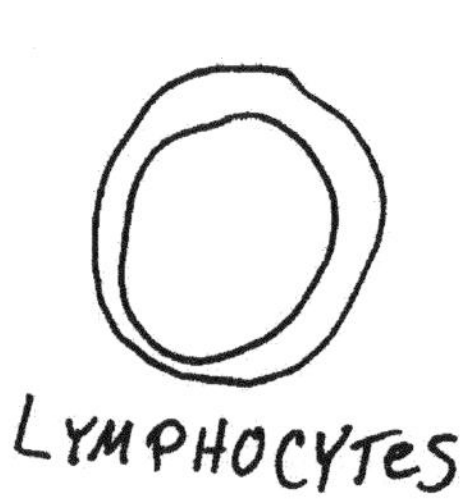

There are three types of lymphocytes: B-cells, T-cells, and natural killer cells.

B-cells produce antibodies that bind to pathogens to enable their destruction.

T-cells coordinate the immune response and defend against intracellular bacteria. The T-cell count is a main indicator of immune system activity in people with HIV/AIDS.

Natural killer cells are able to kill cells of the body, which are displaying a signal to kill them, such as cells infected by a virus or cancerous cells.

Monocytes

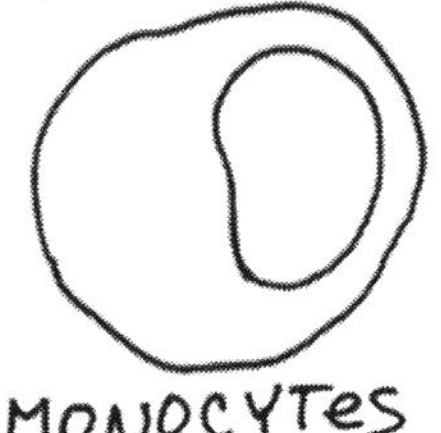

Monocytes also defend against infections, as well as create an antibody response and remove dead cell debris. Once they have moved from the bloodstream into the body tissues, they become macrophages.

Platelets

Platelets, or thrombocytes, are involved in the process of blood clotting. Too few platelets can cause excessive bleeding. If there are too many platelets, blood clots can block blood vessels, causing a heart attack or stroke.

Laws of the Heart

Stress kills: Undo sympathetic stimulation. Inappropriate accommodation to outer environment due to insensitivity to the sensory data.

Any gap between who you are and who you project you are will cause inner insecurity and outward stress.

Sleep rules: Less sleep = more cortisol. More cortisol = more stress.

Aerobic challenge is good training for your heart muscle.

Wonderful nitric oxide: Breath of Fire helps the body facilitate the production of nitric oxide, which aids in good circulation.

Glial cells have been found in the heart. Why?

The plumbing and the blood: Know "The Science of Love." What happens when you relate to the chemicals of love?

A Glimpse at the Relationship to Other Systems

Blood circulation brings nutrition and oxygen to all other organs and the lymphatic system.

The movement of the diaphragm assists circulation.

The vagus nerve of the parasympathetic nervous system and the cardiac nerve of the sympathetic nervous system affect the heart. Parasympathetic stimulus slows the heart; sympathetic stimulus increases the heart rate. Stress is an inappropriate response from the parasympathetic nervous system.

Hormones from the glandular system are carried through the blood.

Cardiopulmonary system is defined by the interaction of the heart and lungs. The breath can change the heart pace through special receptors that monitor blood gas levels and adjust the cardiac rhythm (heart rate).

Negative Influences for This System

- Diets high in (bad) fats (animal fats and saturated fats)
- Obesity
- Lack of movement
- Bad posture
- Nicotine
- Stress
- Alcohol

Some Good Foods for This System

Wheat germ, essential fatty acids, alfalfa, buckwheat, celery, lecithin, green leafy vegetables, nuts, brewers yeast, peas, beans, milk products, cayenne pepper, oranges, lemons, strawberries, calcium, vitamin E, B vitamins, selenium, ginseng, chlorophyll

For blood purification: Yogi Tea, raw onions, grapes

To avoid anemia try: almond milk, raisins, tofu, mung beans, sprouts, tomatoes

How Yoga Helps

- Stress-induced high blood pressure usually has its source in mental, emotional, or physical over-stimulation of the sympathetic nervous system. This causes an increase in the heart rate, strength of contraction, and increased resistance to blood flow to the extremities. Relaxation and effective strategies for coping with unwanted stress can help reduce high blood pressure by lessening the pressures of the venous system and increasing the heart's efficiency.
- Increases oxygen to tissues.
- Aids in weight reduction.
- Strengthens the heart muscles, making each contraction more efficient.
- Increases the amount of small active blood vessels.
- Yoga for cardiac recovery patients, and general heart fitness, are separate areas of yoga training.

Yoga That Affects the Cardiovascular System

- Many of the more aerobic sets (i.e., Abdominal Strengthening p. 57, Sadhana Guidelines,)
- Yoga Running with punching arms
- Frogs, Crow Squats and Chair Pose
- Heart Center sets
- Pranayam, particularly Breath of Fire and Long Deep Breathing
- **For high blood pressure:** breathe through left nostril for 5-11 minutes.
- **For low blood pressure:** breathe through right nostril for at least 5 minutes.
- Arm, torso, and leg lift exercises, such as ego eradicator, cross-crawl type activities, and spinal twists, assist the return of blood and lymph to the heart, and the elimination of excess fluids.
- **Hydrotherapy/cold showers:** The cold shower is known as ishnaan in the East and hydrotherapy in the West. It involves bathing or showering in cold water to open up the capillaries and flush the system. The body is exposed part by part to the icy cold water. The initial response of vasoconstriction is soon replaced by vasodilation. This dilation of the capillaries brings heat to the surface of the skin, causing the skin to become red or flush. Hydrotherapy allows excess heat to radiate from the body and the blood to be infused with oxygen. The flushing of the capillaries may increase circulation by relaxing the capillary system and removing any small restrictions in the capillaries, which are the smallest blood vessels of the body. After the cold water, when the yogi or yogini covers up and meditates, the rush of fresh, red, very oxygenated blood goes deep into the internal organs, rejuvenating the cells of the major organs and tissues.
- **Forget the facelift;** take a cold shower every morning.

The heart is the only organ with its own Chakra.
Love is good for the heart. Hate will kill you.

Some Conditions of the Cardiovascular System

- **Heart attack:** Ischemia, lack of blood to heart muscle, can cause pain and death. The cause is often a clogged coronary artery. The most common artery is the left anterior descending.
- **Stroke:** A vascular accident in the brain. Caused from a weakened blood vessel called an aneurysm. Associated with elevated blood pressure.
- **Congestive heart failure:** When the heart has less capacity to pump, and the vascular system begins to fail. Fluid accumulates in the tissues of extremities, like ankles, feet and hands. Gravity slowly begins to dominate the circulatory system. Fluid forms around the heart.
- **Angina:** Chest pain, which sometimes radiates to the arm, jaw and mid-back, caused by coronary artery insufficiency.
- **C.P.R.:** Cardio (heart) Pulmonary (lungs) Resuscitation. A useful certification for all yoga teachers (American Red Cross).

Vocabulary

Aorta: Major systemic artery; arises from the left ventricle of the heart.
Aortic semilunar valve: These valves reside on the left side of the heart and are similar in their structure and mode of attachment to the pulmonary valves of the right side. However, they are considerably larger, thicker, and stronger.
Artery: A vessel that carries blood away from the heart to the capillaries and has a thicker muscle wall than veins. This muscular wall is responsible for consistent blood pressure.
Atrium: A chamber or cavity; in reference to the heart, one chamber is located on the right side and one on the left side of the heart. The superiorly located heart receiving chambers for the blood returning to the heart.
Bicuspid or mitral valve: The valve that guards the left atrioventricular orifice; it has only two flaps.
Capillaries: The smallest of the blood vessels and the site of exchange between the blood and tissue cells.
Chordae tendineae: Attached to the papillary muscles, they tighten when the papillary muscles contract, preventing the valve flaps from entering into the aorta. They are weblike and move in a graceful flowing motion when functioning.
Cuspid (atrioventricular, A.V. node) valves: These valves allow blood to pass freely from the atria to the ventricles when open. The filling of the ventricles closes the valves and prevents a back flow of blood into the atria when the ventricles contract.
Diaphragm: A partly muscular, partly tendinous partition separating the cavities of the chest from the abdominal cavity. It is very important for breathing.

Diastole: The usual rhythmic dilatation of the heart, especially of the ventricles, following each contraction (systole), during which the heart muscle relaxes and the chambers fill with blood.
Endocardium or visceral pericardium: The thin endothelial membrane lining the cavities (interior of the myocardial wall) of the heart.
Endothelium: Simple squamous epithelium consists of only one layer of flat, scale-like cells. Like cells of the lungs.
Epicardium: The outer layer of the heart wall.
Hemorrhoids: Varicose veins of the anus and rectum.
Inferior (lower) vena cava: Returns the blood to the heart from the body areas that are below the heart.
Interventricular septum or interatrial septum: The partition that divides the heart longitudinally.
Myocardium: The bulk of the heart wall — the thick, middle layer of the heart.
Papillary muscle: This muscle contracts and the chordae tendineae tighten to prevent the valve flaps from entering into the aorta.
Pulmonary circuit: This circuit consists of the pulmonary arteries, which carry deoxygenated blood from the right ventricle to the lungs, capillary beds in the lungs and pulmonary veins, two from each lung, which carry oxygenated blood to the left atrium.
Pulmonary or semilunar valve: Reside on the right side of the heart and consists of half-moon-shaped flaps growing out from the lining of the pulmonary artery and great aorta. When these valves are closed, blood fills the spaces between the flaps and the vessel wall. Each flap then looks like a tiny filled bucket. Blood flowing in then pushes the flaps against the blood vessel walls, collapsing the buckets and thereby opening the valves. Closure of these, as of the cuspid valves, prevents back flow and ensures forward flow.
Sinoatrial (S.A. node, pacemaker) node: This node consists of hundreds of cells that are located in the right atrial wall near the opening of the superior vena cava.
Superior (upper) vena cava: Returns blood to the heart from the body areas above the heart.
Systole: The usual rhythmic contraction of the heart, especially of the ventricles, following each dilatation (diastole), during which the blood is driven onward from the chambers.
Tricuspid valve: The valve that guards the right atrioventricular orifice; it has three flaps.
Vagus nerve: The tenth pair of cranial nerves, arising in the medulla oblongata and providing parasympathetic stimulation to the larynx, lungs, heart (causing a slowing of the heart rate), esophagus, and most of the abdominal organs. It is the most important parasympathetic nerve because it sends nerve fibers to many parts of the body and causes a calming effect. It travels through the diaphragm, and is subject to changes in diaphragmatic tension.
Valve: A structure that allows fluid to flow in one direction only.
Vein: A vessel that returns blood from the capillaries toward the heart; veins are thinner than arteries and are more elastic, having the ability to expand and become like reservoirs. Veins have valves to prevent back flow into the extremities.
Varicose veins: Veins that have weakened valves and wall structure, creating a bulging of the tissue and a pooling of fluid.
Ventricles: Major cavities of the heart; paired (one located on the right side and one on the left side of the heart), inferiorly located heart chambers that function as the major blood pumps; they are the "Double Power Pumps" of the heart.

Study Points

1. Trace the blood flow through the heart.
2. Why do we call the heart a double pump?
3. What is an artery and what is a vein?
4. Is the wall of the vein or the artery thicker?
5. Do the arteries or the veins have valves?
6. How does oxygen get into the blood?
7. What effect do the following activities have on the circulatory system?
   a. Yoga running
   b. Breath of Fire
   c. Long Deep Breathing
   d. Root Lock
   e. Ishnan
   f. Left Nostril Breathing
   g. Hands locked behind heart, one over, one under
   h. Arms up 60%
   i. Stress
   j. Nutmeg, with banana and apple
8. What foods are good for the heart?
9. What type of yoga should a student do if he or she just had heart surgery?

Chapter Five

Respiratory System

Basic Structure/Function

Due to the nature of the human condition, we are oxygen dependent. Our primary access to oxygen is the respiratory system. The physiology of the respiratory system has a profound effect on every system of the body.

The lungs are the largest organs of the body. They are not muscles, just sacs. The diaphragm is a muscular sheet separating the thoracic cavity from the abdominal cavity, which draws fresh air into the lungs and forces old air out. The transfer of oxygen from the air to the blood, and of carbon dioxide from the blood to the atmosphere, occurs in the lungs.

Organs in This System

Lungs

The lungs perform two functions: air (oxygen) distribution to tissues and gas exchange. Air distribution to the alveoli is the function of the tubes of the bronchial tree. Gas exchange between air and blood is the joint function of the alveoli and the networks of blood capillaries that envelop them. The iron (heme-) part of blood bonds with oxygen, which is used in all tissue metabolisms and sustains life.

The lungs also play an important role in the acid-base balance of the blood and tissues. Acidity and alkalinity can disturb mental function and cause nervous system malfunction. Acidity in the system creates an environment for disease organisms and can also compromise the immune system.

The nose acts like a natural regulator, allowing the correct mixture of oxygen and carbon dioxide. When yoga students breathe rapidly through the mouth they may hyperventilate and overload the system with oxygen. The symptoms of hyperventilation are dizziness, numbness around the lips, and hearing a muffled buzzing noise. This can occur in situations of stress and fear. The remedy is to have the person breathe slowly, in and out into a small paper sack. This helps reestablish the correct acid-base balance.

The lungs are controlled automatically by the parasympathetic nervous system through cranial nerve X, the vagus nerve, which initiates bronchoconstriction (tightening of the small air passages). Sympathetic nerves from the thoracic sympathetic chain ganglia also function here, resulting in bronchodilation, preparing the body for action.

The respiratory system functions as an air distributor and gas exchanger in order that oxygen may be supplied to and carbon dioxide removed from the body's cells. In order for air to exchange gases with the blood, air must first exchange gases with blood, blood must circulate, and finally, blood and cells must exchange gases. These events require the functioning of the respiratory system and the circulation system. All parts of the respiratory system – except its microscopic-sized sacs called alveoli – function as air distributors. The alveoli are the only parts that serve as gas exchangers.

Nose

Before the breath reaches the tiny alveolar sacs in the lungs, it is warmed, moistened, and filtered by the nose, trachea, and bronchi. The nose has an internal and an external portion. The external portion is the part that protrudes from the face. This part is considerably smaller than the internal portion, which lies over the roof of the mouth. The interior of the nose, the nasal cavity, is hollow and the septum partitions this into a right and a left cavity. The nasal cavities are separated from the mouth cavity by the palatine bones, which form both the floor of the nose and the roof of the mouth.

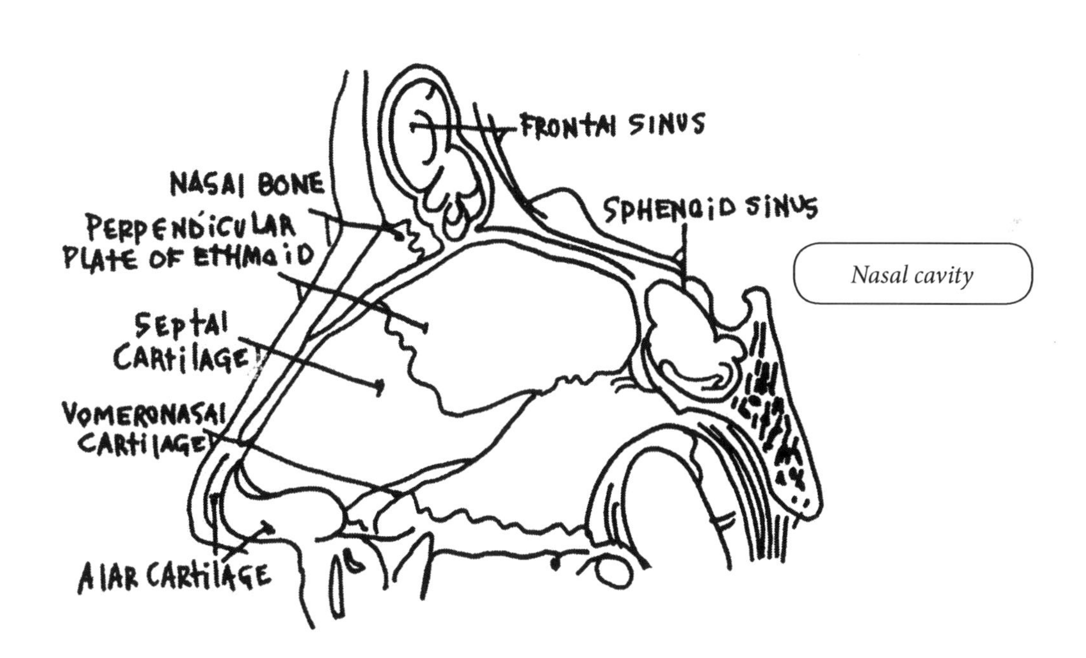

The nose is the passageway for air moving to and from the lungs, filtering the air of impurities, also warming, moistening, and chemically examining it for substances that could be irritating to the mucus lining of the respiratory tract. It is the seat of the olfactory (smell) receptor and it serves as a resonating chamber for speech.

As air passes through the nose it creates a pattern of turbulence, stimulating the brain in specific patterns. The brain functions best when the breath is brought through the nose. The mucus membrane of the nasal cavity facilitates the exchange of gases and substances from the nasal cavity, to the bloodstream.

The nose functions to smell via cranial nerve I, the olfactory nerve, and is also under the control of autonomic (automatic) nerves. Divisions of cranial nerve V, the trigeminal nerve, convey general sensation to the nasal cavity and face.

Pharynx

The pharynx is the hallway for the respiratory and digestive tracts, the passageway that both air and food must pass through before reaching the appropriate tubes. It also has an important role in phonation (the production of sound). As an example, when the pharynx changes its shape different vowel sounds are formed.

The pharyngeal plexus, derived from cranial nerves IX, X and XI, innervates the pharynx.

Larynx

The larynx functions in respiration, since it constitutes part of the vital airway to the lungs. It protects the airway against the entrance of solids or liquids during swallowing and also serves as the organ of voice production by moving air across the vocal cords – a.k.a. the voice box.

The larynx is controlled by the vagus nerve, cranial nerve X.

Trachea

The trachea is a simple and vital organ. It furnishes part of the open passageway through which outside air can reach the lungs. Obstruction of this airway for even a few minutes causes death from asphyxiation.

Air travels through the trachea into bronchi, into bronchioles, ending in thousands of alveolar sacs within the lungs. These sacs are surrounded by capillaries, which move oxygen into the circulatory system and carbon dioxide out. The physiological trigger for respiration is the buildup of carbon dioxide.

The tubes composing the bronchial tree perform the same function as the trachea, to distribute air to the lung's interior.

The alveoli, enveloped as they are by networks of capillaries, accomplish the lungs' main and vital function of gas exchange between the air and the blood. The lung passageways all serve the alveoli just as the circulatory system serves the capillaries.

Certain diseases, pneumonia for example, may block the passage of air through the bronchioles or alveoli. This occurs because inflammation and the accompanying waste plug up these minute air spaces, making the affected part solid. Recovery from this depends largely on the extent of the solidification.

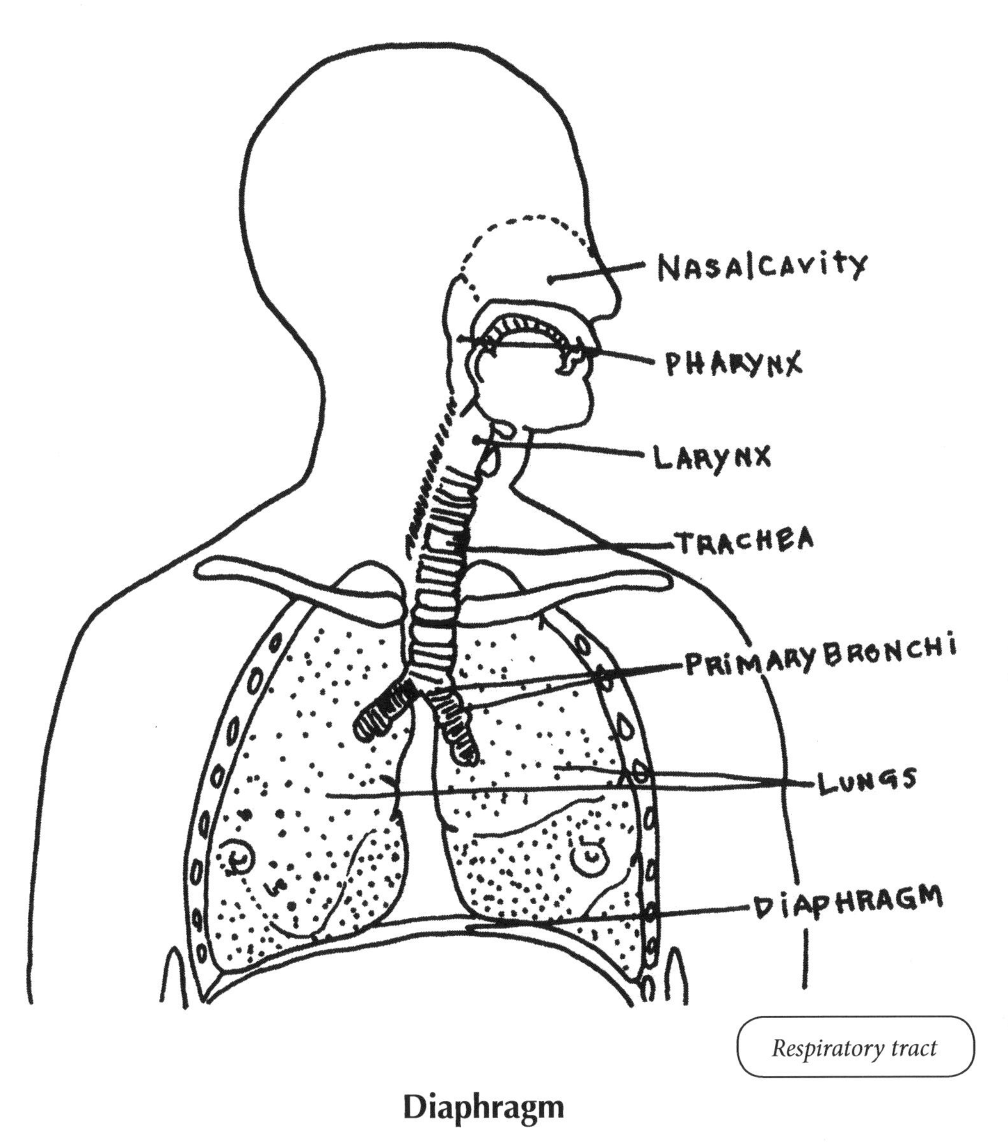

Respiratory tract

Diaphragm

The process of inhalation occurs by a downward, flattening movement of the diaphragm. The external intercostal muscles contract and the ribs lift and expand. This causes an increase in the volume of the chest cavity and a decrease in internal pressure that causes air pressure outside the body to drive air into the collapsed balloons of the lungs through the trachea. The abdominal muscles are relaxed during the inhalation and pulling during the exhalation phase. Exhalation is the elastic recoil of the diaphragm as it relaxes back to its natural dome shape. This pushes the lungs upward as the internal intercostal muscles contract, closing the rib cage and pushing the air out.

The Diaphragm is a domed shape muscle with a central tendon, a costal portion and a crural portion.

The costal portion of the Diaphragm is largest with its fibers radiating out from the central tendon and attaching to the last two ribs, the eleventh and twelfth, laterally and to the sternum in the front.

All the muscle fibers are aligned towards the center of the diaphragm where there is a noncontractile membrane, the central tendon. The central tendon forms the top surface of the dome, which floats freely attached only to the costal and crural muscle fibers.

The crural portion is attached to the upper lumbar spine (4th vertabra), psoas muscle and quadratis lumborum.

The psoas muscle, which is a primary hip flexor and postural muscle, controls the motion of the sacroiliac joint.

The diaphragm essentially contracts against itself. The surface of the diaphragm is slippery, accommodating the movement of many large blood vessels and nerves, which pass through the central part of the diaphragm.

The diaphragm is unique in several ways:

It is intimate with the vital organ of the heart, lungs and accommodates the structures both physically and structurally.

The diaphragm separates the thoracic cavity from the abdominal cavity.

The diaphragm's central tendon is the only tendon not firmly attached to a skeletal structure. It's central tendon pulls the diaphragm down, increasing the vertical volume accommodating inhalation. The heart's pericardium is attached to the central tendon by ligaments. The heart moves up and down on the diaphragm with the breath.

The crural region connects the central tendon to the spine, also some of the primary postural and structural muscles.

The breath influences posture in many ways, both directly by accommodating psoas and indirectly by influencing the structural shape of the ribcage and thirdly through the respiratory signature, which is determined by the balance of the nervous system.

During inhalation, the diaphragm displaces inferior, increasing the vertical diameter of the thoracic cage. The muscles of the abdomen contain the contents of the abdomen and prevent the central tendon from being unstable.

"The more stable the navel point, the more powerful the breath." - DrYogi

The antagonistic and synergistic actions of the diaphragm and the abdominal muscles work together to form a stable structure for breathing. During inspiration the muscle tone of the diaphragm increases, and visa versa during exaltation. Both muscles are contracted during both activities, yet function in a dynamic equilibrium of give and take to accommodate the breath of life.

The somatic and the autonomic nervous systems both participate in breathing. The phrenic nerve controls relaxed breathing and the intercostal nerves activate when labored breathing happens. The phrenic nerve stems from nerves emerging from cervical vertebrae 3, 4 and 5 as the cervical plexus. Relaxed breathing from the diaphragm is directed by the cell bodies located in the upper cervical spine. The cell bodies of the intercostals nerves are located in the thoracic spine (T-1 to T-12).

Breathing continues either consciously under the control of the cerebral cortex or automatically under control of the pons and the medulla in the brainstem. The somatic motor circuits for respiration delicately balance the tone of the breath, either smooth, gasping, deep or shallow, synchronized or disharmonious.

Smooth, full breathing is an exquisite action of functional beauty and has an effect on all the trillion or so cells in the body.

The breath is the commander of the nervous system and can synchronize the physiology into a symphony of health and abundance. Developing your relationship with your breath, guided by a calm mind, can greatly further your meditative experience.

The motion and function of the diaphragm profoundly affect the posture, the physiology, the mental attitude, and the projection of an individual. It defines your shape and is the vital intersection between mind and body.

"As you breathe, so shall you live." - DrYogi

A Glimpse at the Relationship to Other Systems

The efficiency of the respiratory system has an effect on all other systems. The root of ALL death is a lack of oxygen.

Proper breathing massages the internal organs, promotes good digestion, and nurtures a peaceful, tranquil and alert attitude.

The patterns of breathing can develop tendencies of the nervous system. A long deep breath can access the parasympathetic nervous system and a rapid, shallow breath can evoke the sympathetic nervous system. This balance of the nervous system affects everything from digestion to sex.

Breathing creates movement, a rhythmic pulse, which elongates the spine, and facilitates the pumping of spinal fluid up and down the spine.

The interaction of the heart and lungs — the Cardiopulmonary System — is significant. Patterns of the heart's rhythm can be changed by breath, which affects the vagus nerve.

Erratic breath is associated with sickness.

Negative Influences for This System

- Smoking decreases the ability of the lungs to cleanse themselves of toxins and particles. Smoking diminishes cilia activity and ultimately destroys the cilia. When this function is lost, coughing is the only means of preventing mucus from accumulating in the lungs. For this reason, a smoker with respiratory congestion shouldn't be given medications that inhibit the cough reflex. These particles remain and create disease, especially cancer.
- Eating an oil-free diet can dangerously affect the lungs, which need to be lubricated with a substance called surfactant (this is a rare occurrence).

- A chronic habit of poor, inefficient breathing can create excess abdominal tension, stress, and digestive difficulties.
- A deviated nasal septum can cause an ida or pingala flow imbalance.
- A pingala (right nasal) block can create a lack of projection, poor digestion, and decreased clarity of thought.
- An ida (left nasal) block can create a lack of the ability to sleep, listen, and/or relax.
- Fear can manifest in diaphragmatic tension.

Some Good Foods for This System

- Oranges
- Raw fruits and vegetables (especially carrots, beets, and radishes)
- All berries
- Nuts (especially walnuts)
- Millet
- Ginger, ginger tea
- Clean water
- Black pepper corn tea

How Yoga Helps

Long Deep Breathing

- Balances the parasympathetic and sympathetic nervous systems, creating a deep, effective relaxation by slowing down the mental intellect.
- Creates a pulsing, which stimulates the pituitary, or master gland, allowing the development of intuitive consciousness.
- Pumps the cerebrospinal fluid to the brain.
- Reduces toxic buildup in the lungs, by properly cleansing the alveolar sacs.
- Stimulates the production of chemicals in the brain which work against depression.
- Cleanses the blood.
- Increases the available amount of oxygen and prana.
- Helps to release blockages in the flow of energy through the body (nadis).
- Activates and cleanses the nadis (nerve channels).
- Activates the physical and emotional healing processes.
- Breaks unconscious patterns and addictions.
- Facilitates the healthy stretching of the alveoli and strengthens the diaphragm.

Breath of Fire

- Maximizes the blood chemistry making it more alkaline, a desirable state for calming of the mind.
- Activates the vagus nerve and balances the autonomic nervous system.
- Cleanses the blood by stimulating the process of removing toxins and deposits from the cells, lungs, and mucus membranes.
- Stimulates the solar plexus, freeing the natural flow of energy, allowing it to flow through the body.
- Increases the capacity of the nervous system to withstand stress.
- Produces an alpha rhythm in the brain.

- Increases physical stamina.
- Stimulates the navel point, strengthening the nervous plexus or nabhi.
- Helps to break addictions.
- Increases the lung capacity, resulting in better health and more enjoyment of life.
- Helps to balance the brain.
- Charges the electromagnetic field.

Yoga That Affects the Respiratory System

- All Kundalini Yoga sets, exercises, pranayama, and meditations increase the effectiveness of the respiratory system
- New Lungs and Circulation (also called Something New! Yoga for the 80's, p. 32-33)
- Preparatory Exercises for Lungs, Magnetic Field and Meditation (p. 85-86, Sadhana Guidelines)
- Alternate nostril breathing
- Three minutes of Breath of Fire
- Any yoga that develops navel point strength and stability
- Walking and Breathwalk

Laws of Lungs and The Breath of Life

The consciousness of breathing. Breath is a unique function; it is often unconscious and yet can be overridden by a conscious command. Breath is one of the first pathways of consciousness. When doing most Kundalini Yoga exercises the breath is very important. Instructing students on how to breath with maximum lung volume, as well as total exhale, is critical for them to have a maximum neuro-glandular experience. The control of the breath is the first step in self-guidance of the mind. "If you can control your breath, you can control your mind, and if you control the mind you can control the world." Focusing on the breath is often the first pathway toward awareness.

The kinetic physiology of breathing. With each breath all the internal organs dance and move by the kinetic energy of the air moving in and out of the lungs. As air enters the lungs, the diaphragm muscle drops into the abdominal cavity. This begins a cascade of events that massages and facilitates the function of all the internal organs. The kidneys move laterally, the intestines, all 22 feet, move in an orchestrated manner that encourages the movement of food through the body (motility). The heart is squeezed and the liver moves in rotation on its own axis. If every breath is full, the organs move naturally in rhythm and synchronization.

The navel point and the diaphragm. There is a dynamic dance between the navel point and the major breathing muscle, the diaphragm. Doei Shabd Kriya is a good meditation that illustrates this relationship. Most of the time the neuro wiring that tells the navel point to contract also tells the navel point to contract. Learning through practice can give you the ability to move them separately at will. This type of focus and skill opens the door to the sensitivity and inner control that gives yogis the ability to control (guide) their physiology by command. The power of the navel point fuels the motion of the diaphragm, and the diaphragm moves the entire body and leads the psyche to the mantra. Breath awareness naturally leads to navel point awareness. All good posture begins with the complete conscious breath and the symmetry and balance in the navel point.

The Nose Knows! Use the nose, not the mouth for breathing. It is your natural, built-in, never-fail regulator. The membranes in the nose absorb oxygen and warm the air entering the respiratory system. The small hairs filter out particulate matter (dust, dust mites, other insect parts and pollen) in the air. The exhale is also through the NOSE. Unless otherwise instructed, ALL breathing is through the nose! When the nose controls exhaling, the pressures in the Respiratory System are more controlled by the consciousness (nervous system) and less by automatic glandular reactions. When you breathe through the nose there is a natural switching of the nostrils between left and right every 2.5 hours or so. Some specific breaths require heavy mouth breathing; they are more associated with a glandular experience. By breathing through your nose, you may find that the amount of unnecessary drama in your life is significantly less.

The Sacred Breath. We begin our time on earth with a deep inhale, and we leave with our last exhale. Our breath is truly our vital force. The currency of all life is the breath. The lung tissue, if opened up, has the surface area of a tennis court. That is a lot of surface, all of which can be exposed to the beautiful ethers of life, and the many environmental irritants we may be exposed to. When we are given the gift of life, it is quantified by an amount of prana. The ethereal currency of prana is converted to the local currency of breaths during life. Yogis use specific techniques to utilize the breath efficiently. Yogis value the breath as a gift, to be cherished and valued. Breath of Fire and Long Deep Breathing are examples of breath awareness techniques.

Some Conditions of the Respiratory System

Asthma: A condition where there is a constriction of the bronchial airways.

COPD: Chronic Obstructive Pulmonary Disease; refers to a group of diseases that all create some recurrent obstruction to airflow within the lungs.

Emphysema: Abnormal enlargement of the little sacs that make up the lungs. It is similar to when a person turns a 22-room house into an open loft.

Pneumonia: An acute lung infection that fills the lung with fluid.

Vocabulary

Alveoli: The final subdivision of the respiratory tree in the lungs.

Alveolar ventilation: The volume of inspired air that actually reaches alveoli; equal to tidal air volume minus dead space air volume.

Anatomical dead space: The volume of air that fills the nose, pharynx, larynx, trachea, bronchi, and smaller tubes but does not descend into alveoli, therefore takes no part in gas exchange.

Diaphragm: A membrane or partition that separates one thing from another; the muscular partition between the thorax and the abdomen.

Expiratory Reserve Volume (ERV): The additional amount of air that can be forcibly expired after a normal inspiration and expiration.

Inspiratory Reserve Volume (IRV): The amount of air that can be forcibly inspired after normal inspiration; measured by having the individual expire normally after forced inspiration.

Larynx (voice box): Lies between the root of the tongue and the upper end of the trachea just below and in front of the lowest part of the pharynx. Its main two tasks are to provide a permanently open airway and to act as switching mechanism to route air and food into the proper channels. Its third function is voice production.

Lungs: The two sponge-like respiratory organs in the thorax that oxygenate the blood and remove carbon dioxide from it.

Minimal Volume (MV): The trapped air that remains in the lungs after the residual volume is eliminated (as in pneumothorax/collapsed lung).

Nasal cavity: This hollow cavity constructed mostly of bone and hyaline cartilage in the 1/10 interior portion of the nose.

Pharynx: Commonly called the throat, it is a funnel-shaped passageway that extends from the base of the skull to the level of the sixth cervical vertebra and serves as a common passageway for food and air.

Primary bronchi: Either of the two main branches of the trachea, or windpipe. The trachea divides at its lower end into two primary bronchi.

Study Points

1. Which lung has three lobes, and which has two lobes?
2. How does pranayam affect the respiratory system?
3. How can a person increase the efficiency of their lungs?
4. How can a person exercise the diaphragm muscle?

CHAPTER SIX

THE SKELETAL SYSTEM

The Skeletal System is the structure that retains the form of the human body. Without bones we would be a blob. Joints provide flexible connections between your bones. The Skeletal System also protects the internal organs and provides attachment sites for the organs. There are 206 bones in the adult human skeleton. It works closely together with the muscular system to allow us to move.

BASIC STRUCTURE/FUNCTION

The Skeleton

The skeleton has two general divisions:

Axial Skeleton

Primarily consists of supportive structures of the body and includes:

- skull
- Spine
- Sternum
- Ribs
- Hyoid bone

Appendicular Skeleton

Primarily provides mobility for the upper and lower limbs and includes:

- Pectoral girdle
- Pelvic girdle
- Arms
- Hands
- Legs
- Feet

Fractures and dislocations are more common in the appendicular skeleton but more serious in the axial skeleton.

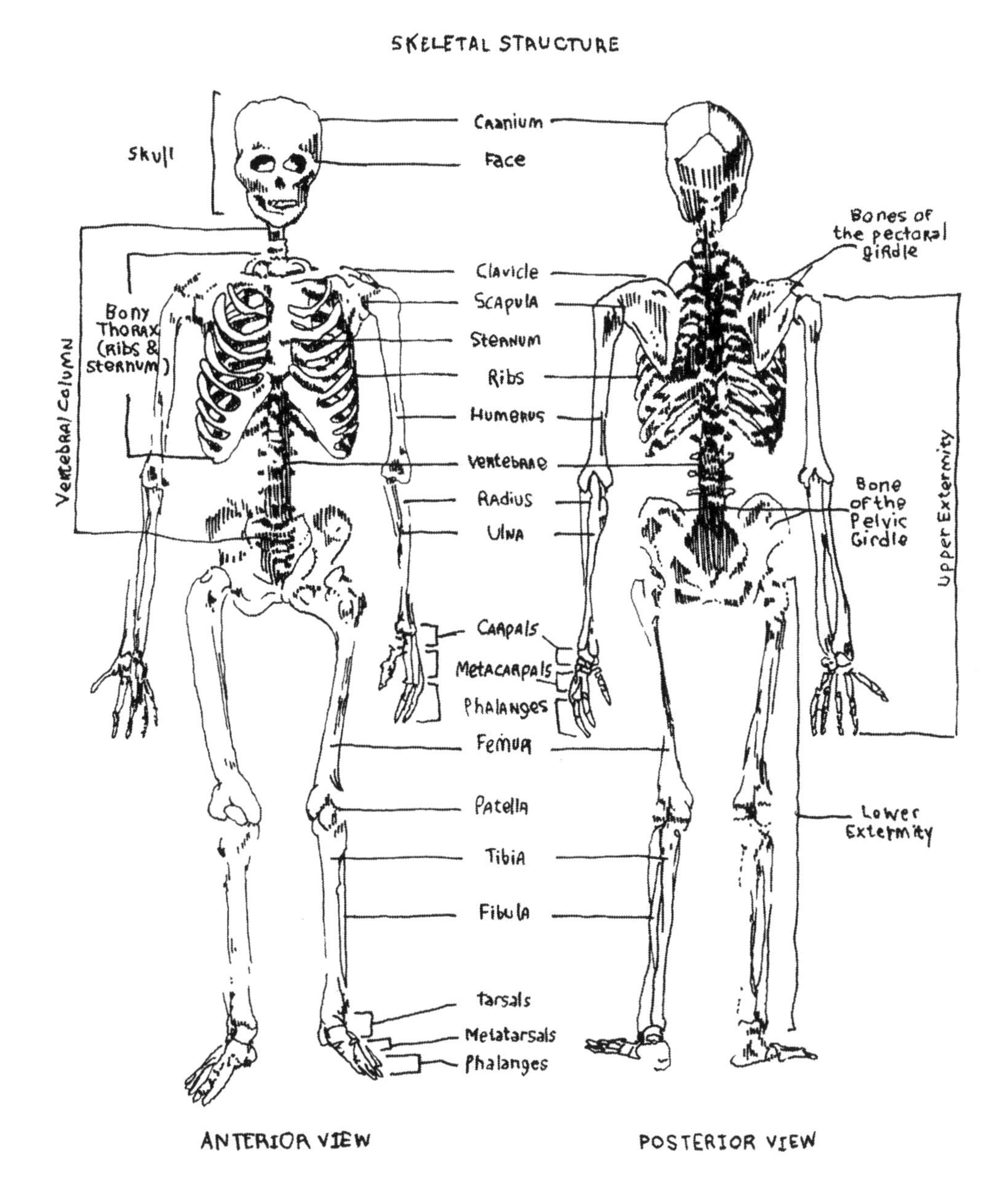

Organs in This System

Bones

Bones are a major organ of the Skeletal System; their purpose is to protect and support the function of other organs in the body. Bones store minerals and provide blood protection while producing red and white blood cells. Bones are comprised of an intricate meshwork of fibers and cells that are impregnated with calcium salts.

Parts of a Bone

Epiphysis: The end of a long bone where bone growth occurs. The epiphyseal line or epiphysis plate can be damaged during youth.

Diaphysis: The shaft of a bone. It consists largely of compact bone, in tubular format, creating the strongest structure for the material.

Medullary cavity: The cavity deep inside the diaphysis (the shaft). It serves as a space for bone marrow. In adult humans, blood cell production occurs in this bone marrow.

Periosteum: A fibrous sheath that surrounds all bones. The peri- (around) osteumis richly innervated with nerve fibers, which is why a broken or bruised bone causes so much pain.

Articular cartilage: A smooth, white tissue that covers the ends of bones where they come together to form joints. It allows the bones to glide over each other with very little friction. As children, our bones start out as cartilage templates and then slowly fill in with calcium as we grow older. The last of these endplates to grow together are in the hip region and do not fuse until the age of 13.

Types of Bones

Long bones: Upper and lower arms and legs, fingers and toes

Short bones: Wrist and ankle bones

Flat bones: Ribs, scapula and certain bones (frontal, parietal) of the cranium (skull)

Irregular bones: Bones of the spine (vertebrae, sphenoid) and bones of the internal skull

Sesamoid bones: Bones embedded in tendons, like the patella and pisiform

Purpose of Bones

Support: The bones are the supportive framework of the body. An extensive network of nerves around bones and joints gives bone the lowest threshold for pain out of any of the deep body tissues.

Protection: Hard bones protect the soft organs. The skull protects the brain and the ribcage protects the lungs and the heart.

Movement: Bones and the muscles create levers.

Calcium Reservoir: Bones contain a supply of calcium(and several other minerals)maintaining homeostasis of blood calcium.

Hematopoiesis (Blood Production): Large bones, like the femur, contain red marrow, which makes new red blood cells in the adult human. The bodies of the vertebrae, the sternum, ribs, cranial bones, femur, and humerus all produce new red blood cells.

It is a part of the ancient technology of Kundalini Yoga to keep the temperature in the thigh/femur area buffered from rapid temperature change with the use of kacheras. Kacheras are a special type of shorts worn to buffer the thighbone (femur) and pelvis from intense temperature fluctuations (boxer shorts can also be worn).

Cartilage

Cartilage is a connective tissue located throughout the body. It is composed of important cells and fibers. Cartilage supports the structure of many body parts and functions like the bones, ribs, ears and nose.

Ligaments

Ligaments are fibrous tissues that provide stability to joints throughout the body during rest and active movement. Ligaments connect bone-to-bone.

Tendon

Tendons provide adhesive strength as they connect muscle to bone. Tendons withstand pressure and tension; they work collectively with muscles to enable rigorous movement.

MUSCLES AND TENDONS MOVE THE BONES, WHILE LIGAMENTS RESTRICT THE MOVEMENT, PROTECTING THE ARTICULATIONS AND THE NERVOUS SYSTEM.

Bursae

A bursa (plural bursae) is a small fluid-filled sac lined by synovial membrane with an inner capillary layer of slimy fluid. It provides a cushion between bones and tendons and/or muscles around a joint. This helps to reduce friction from adjacent articulating surfaces and allows free movement. These are found around most major joints of the body. They are common in sites where muscles, skin, ligaments, or muscle and bone override. Most bursae are present at birth, but additional ones may develop in sites where excessive motion occurs.

Joints

Joints are mainly classified structurally and functionally.

Structural classification is determined by how the bones connect to each other, while **functional classification** is determined by the degree of movement between the articulating bones. There is significant overlap between the two types of classifications.

There are three structural classifications of joints:

Fibrous joint: Joined by fibrous connective tissue
Cartilaginous joint: Joined by cartilage
Synovial joint: Not directly joined

Functional classification focuses on the amount of momentum allowed by the joint.

There are three functional classifications of joints:

Synarthrosis: Immovable joint. Most synarthrosis joints are fibrous joints like skull sutures.
Amphiarthrosis: Slightly movable joint. Most amphiarthrosis joints are cartilaginous joints like vertebrae.
Diarthrosis: Freely movable joint. All diarthrosis joints are synovial joints, like the shoulder, hip, elbow, and knee.

Fibrous Joints

Fibrous tissues join fibrous joints; there is no joint cavity present. Some fibrous joints are slightly movable. However, most are classified as synarthrotic and essentially move very little. There are three basic kinds of fibrous joints.

The first type of these joints are sutures, which occur between the bones of the skull and have wavy, interlaced, sharp bone edges that interdigitate by overlapping and underlocking the joint articulation. These fibrous joints are normally more moveable in youth and more rigid in adulthood.

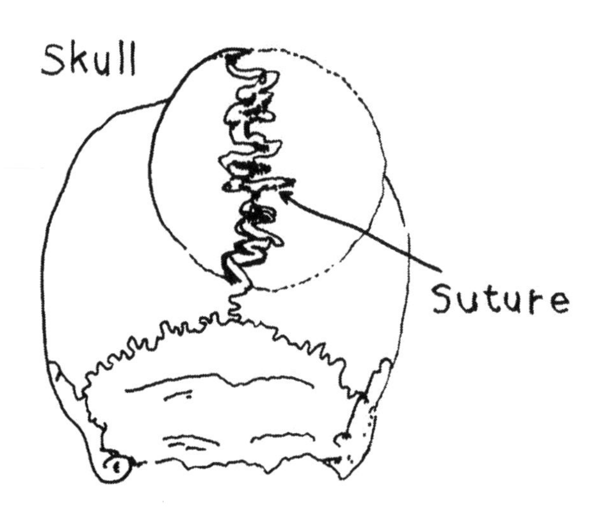

Second is the syndesmosis, a fibrous joint in which the bones are connected on a sheet or cord of fibrous tissue called a ligament, or interosseous membrane. These are slightly movable joints. An example is the upper bone of the leg where the fibula and the tibia come together. There is a short thick fibrous ligament holding them in place.

The third type is called a gomphosis, which is from the Greek word gompho, meaning "nail" or "bolt" and is referred to when describing the manner in which the teeth are embedded in their sockets.

Cartilaginous Joints

Cartilaginous joints are linked by cartilage. They lack a joint cavity.

The two types of cartilaginous joints are synchondroses and symphyses.

The epiphysial plate in a young person is a synchondrosis; two bones joined with cartilage and no cavity.

The second type of cartilaginous joint is a symphysis. This type of joint uses fibrocartilage as a compressible, reliant shock absorber that limits the amount of movement at that joint. Examples include the intervertebral joints and the pubic symphysis.

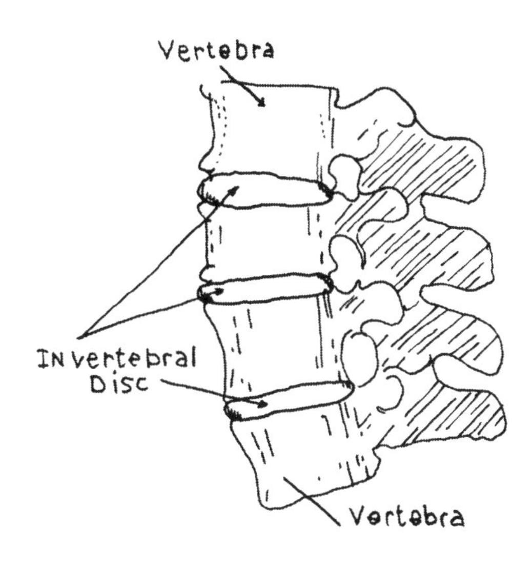

Synovial Joints

Synovial joints are those joints whose articulating surfaces are separated by a fluid-containing joint cavity. This architecture allows substantial freedom of movement in all joints of the limbs. Most joints of the body fall into this class.

Synovial joints typically have the following distinguishing features:

Surfaces of the bones covered with a glass-smooth articulated cartilage known as hyaline cartilage.

An articular capsule encloses the entire joint in a double-layered capsule. The outside layer is tough and is known as the fibrous capsule, and is continuous with the periosteum of the bone.

Synovial fluid (Your own lubricating oil) occupies the free space in synovial joints that is created by specialized cells of the synovial membrane secreting hyaluronic acid. The synovial fluid lubricates the joints and prevents the bone ends from touching each other. This is a fine quality lubricant, offering protection friction, as well as nourishing the structures within the joint capsule. Golden Milk helps create good quality synovial fluid. A major component of turmeric is sulfur; sulfur is a key component in synovial fluid.

Reinforcing ligaments, some inside the capsule and some outside the capsule, hold the joint in place.

Six General Types of Synovial Joints:

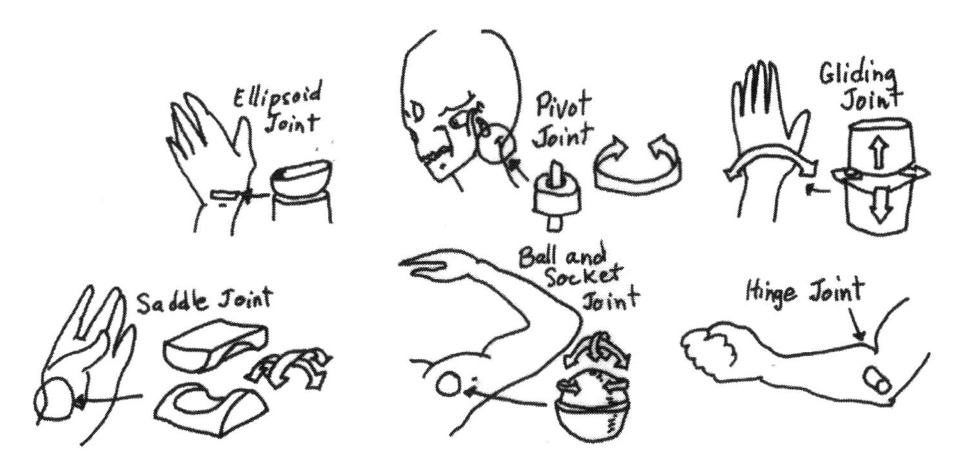

Ellipsoid: The oval-shaped end of one bone (convex surface) articulates with the elliptical basin (concave surface) of another bone. It allows flexion/extension and abduction/adduction as in the radiocarpal joint in the wrist.

Pivot: It allows one bone to rotate around the surface of another bone, as in the rotation of the head.

Gliding: This joint is between two flat or only slightly rounded surfaces and allows bones to glide past each other such as those between the carpal bones in the wrist.

Saddle: This type of joint allows for back and forth and side to side motion but limited rotation. There is a saddle joint between the trapezium (one of the small carpal bones in the wrist) and the first metacarpal bones.

Ball and Socket: These are the most freely movable joints and they are universal in their movements. These are shoulder and hip joints. They provide tremendous opportunity for mobility and injury.

Hinge: This type of joint allows for movement much like that of a door hinge. The knee and ulna part of the elbow are hinge joints.

Joint injuries occur when one or more structures of the joint are physically traumatized through excessive movement outside the limit of anatomical integrity or from an outside traumatic force.

Motions Allowed by Synovial Joints:

Gliding: Smooth movements as in the wrist.

Flexion: A bending momentum that decreases the angle of the articulation and brings the two bone ends closer together.

Extension: The reverse of flexion and involves movement that increases the angle between articulating bones. Examples of flexion and extension occur in both the elbows and knees.

Abduction: Moving away from the midline of the body. An example is raising the arm or leg away from the body.

Adduction: Moving toward the midline of the body.

Circumduction: From the word "circum" (around) and "duco" (to draw), as when you make circular motions, as if drawing a circle, with the arms or legs.

Rotation: The turning movement of a bone around its own long axis; special movements which only occur in specific joints in the body; the description of their motion does not usually apply to other regions. This is the only movement allowed between the first two cervical vertebrae and is common in the shoulder and hip area.

Supination and Pronation: This movement occurs primarily in the wrist and forearm. Supination is the movement of the forearm when the palm faces up towards the sky as if holding soup in your hand. Pronation is the opposite.

Inversion and Eversion: These are specialized terms to refer to movements of the foot. An inversion ankle sprain is one where the ankle is turned in such a way that the sole of the foot is turned towards the middle of the body, with the big toe up and the little toe down. This is the most common type of ankle sprain. An eversion sprain is one where the foot turns laterally away from the body with the big toe down and the little toe up, and is often a more severe type of sprain.

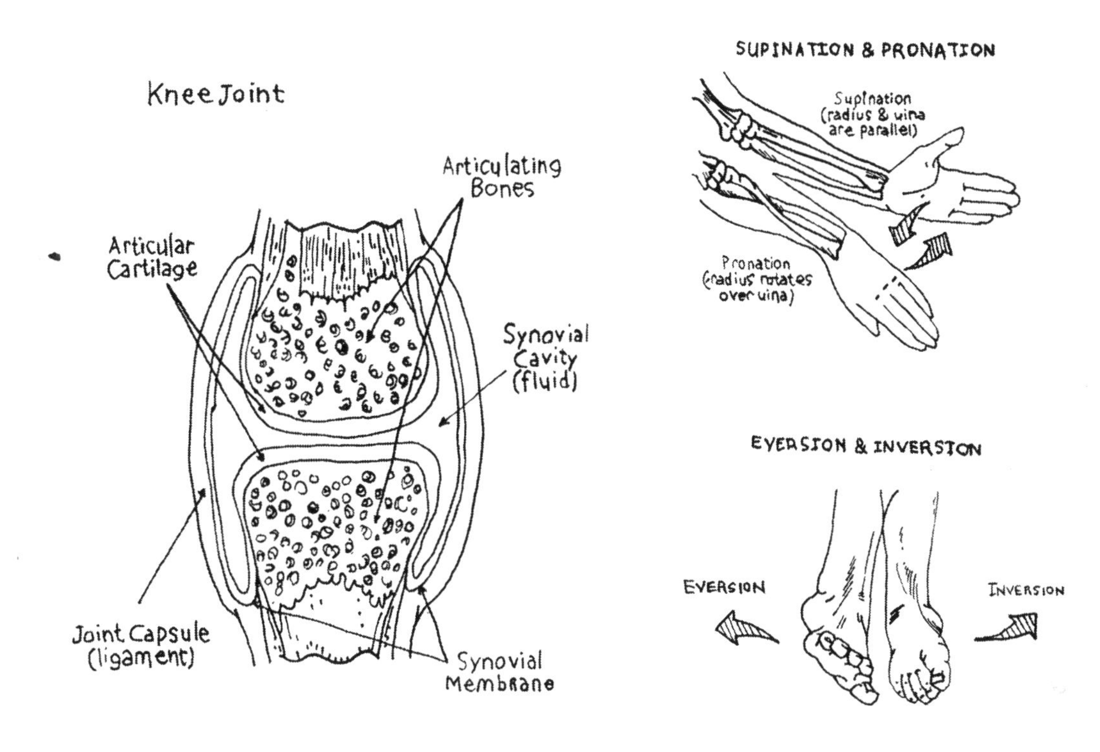

How Yoga Helps

Kundalini yoga helps the Skeletal System in 2 ways:

Weight bearing exercises help bones stay strong by encouraging ostoblastic (Bone making) activity.
Moving joints in a gentle manor, helps nourish the fluids of the joint capsule and maintain the ligaments and tendons.

For an in depth look at musculoskeletal motion read the 13 Principles of Yoga-Motion chapter.

Some Conditions of the Skeletal System

Osteoporosis: This is a condition in which bone reabsorption occurs faster than bone deposition. In other words, you lose bone faster than you make it. Osteoporosis is most common in American and European post-menopausal women. This bone-thinning condition is often the cause of fracturing hip and pelvic bones.

Clubfoot: A common defect, clubfoot twists the sole of the foot towards the middle of the body. There are several causes and methods of correction.

Injuries to Joints

Arthritis: Arthro = joint; itis = inflammation of; Arthritis = inflammation of joints. There are many types of arthritis.

The two major categories of arthritis are osteoarthritis and rheumatoid arthritis.

Osteoarthritis: the most common form of arthritis. It is caused by wear and tear on joint structures creating compensations, such as bone spurs and inflammation of the joint capsule. Every mammal on earth gets osteoarthritis with the exception of two species – the bat and the sloth. Osteoarthritis is aggravated by an increase in gravity (overweight), alteration of normal joint function, and repetitive motion on joint tissues. Joints that are do not move for long periods of time also develop inflammation, fibrosis, and alterations of joint tissue.

Rheumatoid Arthritis: most commonly occurs bilaterally and symmetrically in the body. It is not caused from the overuse of a joint, but is actually an autoimmune disease in which the body crea

Common treatments for this condition are deep heat and rest. Often patients receive great reduction of symptoms by changing to a less inflammatory diet, eliminating dairy products, and generally increasing the function of the immune system. Eating onions, ginger, and garlic is a good start.

Vocabulary

Axial skeleton: The portion of the skeleton that forms the central axis of the body. This includes the bones of the skull, the vertebral column, and the bony thorax.

Appendicular skeleton: The bones of the limbs and limb girdles that are attached to the axial skeleton.

Articular capsule: Double-layered capsule composed of an outer fibrous capsule lined by synovial membrane; encloses the joint cavity of a synovial joint.

Articulation: Joint, where two bones meet.

Bone marrow: The soft, vascular, fatty tissue that fills the cavities of most bones.

Bursa: The fibrous sac lined with synovial membrane and containing synovial fluid; occurs between bones and muscle tendons (or other structures) where it acts to decrease friction during movement.

Carpal: Pertaining to the wrist.

Cartilaginous: Of or like cartilage, gristly.

Clavicle: A bone connecting the sternum with the scapula; collarbone

Diarthrosis: Any articulation, as of the hip, permitting free movement in any direction.

Diaphysis: The shaft of a long bone, as distinguished from the growing ends.

Malleus: The tiny middle ear bone that is shaped like a hammer.

Incus: The central one of the three small bones in the middle ear; it is shaped somewhat like an anvil.

Stapes epiphysis: The innermost of the three small bones in the middle ear; shaped like a stirrup.

Femur: The largest, longest, and heaviest bone in the body, extending from the hip to the knee.

Fibrous: Containing or composed of fibers, threadlike structures.

Fibula: The long, thin outer bone of the human leg between the knee and the ankle.

Humerus: The sole bone of the upper arm.

Hyaline: Any of various glassy translucent substances, especially such a substance occurring normally in vertebrate cartilage.

Joint: A place or part where two things or parts are joined; the way that two things are joined at such a part.

Joint cavity: A unique feature of synovial joints in that it is more of a potential space than a real one because it is filled with synovial fluid.

Mandible: The lower jaw.

Maxilla: The upper jaw, or a major bone or cartilage of the upper jaw.

Metacarpus: The part of the hand consisting of the five bones between the wrist and the fingers.

Radius: The shorter and thicker of the two bones of the forearm on the same side as the thumb.

Ribs: There are twelve pairs of ribs. The upper seven pairs are the true ribs. They are attached to the sternum by costal cartilage. The next five pairs, the false ribs, do not attach to the sternum directly. The last two pairs are called the floating ribs because they do not attach to the sternum at all. The top ten ribs articulate with the transverse process of the vertebrae above and below it. The flexibility of the spine, as well as the shape of the rib cage both directly contribute to the way we breath. The depth and width of our breath has a profound effect upon our entire physiology.

Sacrum: A thick, triangular bone situated near the lower end of the spinal column, where it joins both innominate (not named) bones to form the dorsal part of the pelvis. It is formed of five fused vertebrae.

Scapula: Either of two flat, triangular bones in back of the shoulders.

Sternum: A thin, flat structure of bone and cartilage to which most of the ribs are attached in the front of the chest.

Symphysis: A joint in which the bones are connected by fibrocartilage.

Synovial fluid: The fluid secreted by the synovial membrane; this fluid lubricates joint surfaces and nourishes articular cartilages.

Tibia: The inner and thicker of the two bones of the human leg between the knee and the ankle; shinbone.

Ulna: The larger of the two bones of the forearm of humans, on the side opposite the thumb.

Xiphoid process: Point at the end of the breastbone, shaped like a sword.

Study Points

1. What is bone made of?
2. What are the two major functions of bones?
3. What are the names of the bones in the arm and leg?
4. How many bones are in the head, the hand, the foot, and the spine?

Chapter Seven

The Muscular System

Muscles have only one action – Contraction!

The Muscular System's main role is to provide movement. Muscles work in pairs to move limbs and provide mobility. Muscles also provide strength, posture, balance, and heat for the body to keep warm. Muscles control the movement of materials through some organs, such as the heart and circulatory system and stomach and intestine.

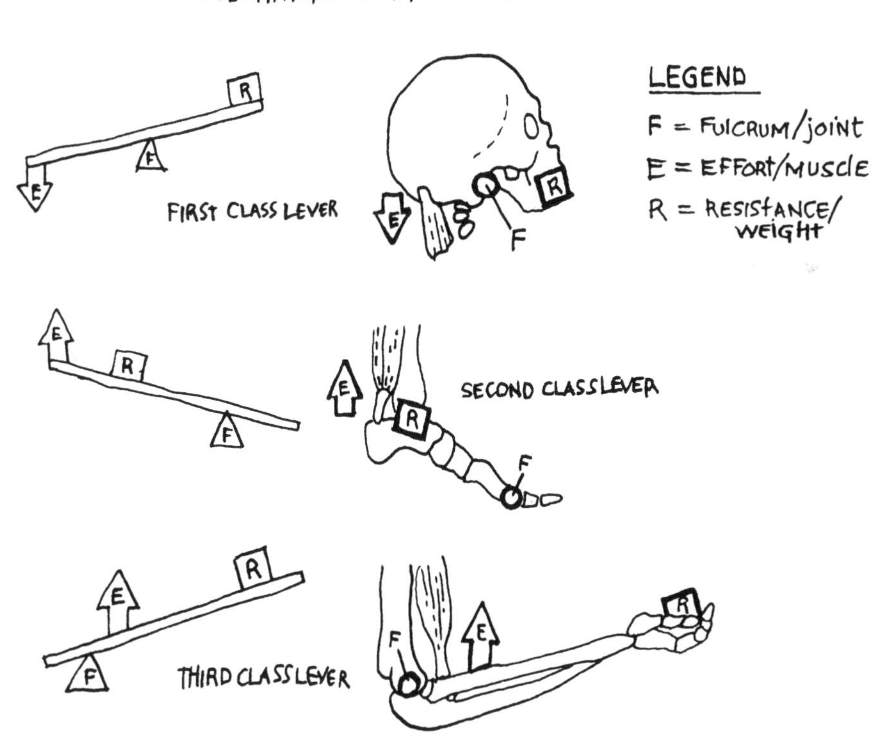

Levers are simple machines employed by the skeletal muscles. The degree of muscular effort required to overcome resistance depends upon the force of resistance (weight) and the relative distances from fulcrum to point of resistance (f-r) and from fulcrum to point of muscular effort (f-e). The position of the fulcrum relative to points (f) and (e) determines the class of the lever system in use.

Basic Structure/Function

Between the bones are joints, which allow for motion. The joints have cartilage on the ends of the bones to allow for smooth motion; ligaments act as hinges for the joint. The muscles are pulled across joints and connect to the bones via tendons. The fixed end is called the origin; the moving end is called the insertion.
Muscles can only pull in the direction of contraction -- pulling the origin toward the insertion. Many injuries are the result of improper controlled elongation; in other words, asking a joint to move in a way it was not designed for will injure the joint.

Mechanics of Movement

The physical advantage that muscles have is due to the location of the fulcrum. The fulcrum is around the joint, the body part that moves is the resistance, and the muscle is where the effort is exerted. The physical advantage that the human form has is due to the placement of the muscles in relation to the load.
First Class Lever: The fulcrum is between the muscle and the resistance.
Second Class Lever: The resistance is in the middle between the muscle and the fulcrum.
Third Class Lever: The muscle is between the fulcrum and the resistance.

Muscle cells have two major components: actin and myosin. Together they form the muscle unit, called a sarcomere. The actin and myosin fibers are bound together by cross bridges. The myosin stays stable and the actin slides toward the myosin, resulting in the shortening of the muscle. Every time a muscle contracts and shortens, these cross bridges move. Every time these cross bridges move, energy (ATP) is used. To keep a muscle under constant contraction uses energy and this can create pain and structural dysfunction.
The single muscle fibers are composed of many sarcomeres. These fibers are bound together in strands and bound by successively larger sheaths until the muscle belly is formed. This muscle would not develop or move without nerve stimuli.
The motor nerve has a special cell at the neuromuscular junction called a Schwann cell. At this neuromuscular junction, the chemical acetylcholine is released, once again creating an electro-biochemical event we call life.
Each nerve and adjacent muscle fiber is referred to as one motor unit. Each nerve commands several fibers. In other locations, one nerve (somatic motor neuron) activates thousands of fibers and, in some locations one nerve stimulates only a few fibers. For example, in the leg thousands of fibers are motivated to move, and in the hand some nerves activate very few muscle fibers.
The peripheral nervous system controls our muscles, which move our limbs and body. There is an outgoing message that is conveyed by an afferent (towards the brain) motor neuron and an incoming message that is conveyed by an efferent (away from the brain) sensory neuron.
In addition to this intricate network of motor-stimulating neurons, there are almost as many sensory neurons feeding information regarding the condition of the muscle, such as tone, pressure, and temperature.

No nerve supply = paralysis.

Types of Muscles

There are three types of muscles: smooth, cardiac, and skeletal.

Smooth Muscle

Smooth muscle is an involuntary non-striated muscle. Smooth muscles are arranged in layers with the fibers in each layer running in a different direction. This makes the muscle contract in all directions. Smooth muscles are found in the hollow parts of the body like the stomach, bladder, intestines, blood vessels and other tube-like structures.

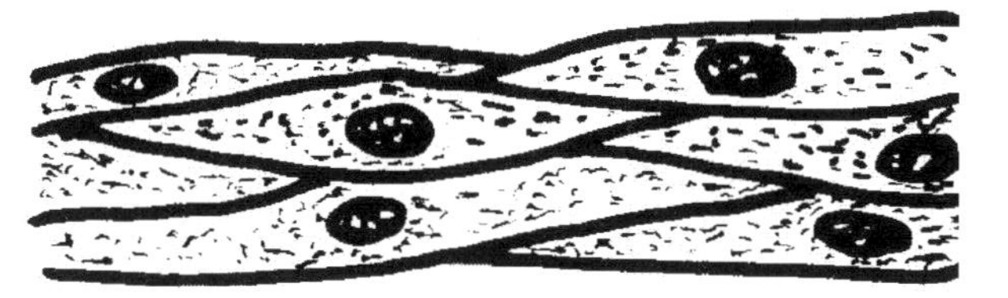

Smooth Muscle

Both arteries and veins have layers of smooth muscle surrounding them. The largest artery, the aorta, also has cardiac muscle fibers in its walls for the first few inches of its length immediately leaving the heart. The circulatory system comprises two major divisions, one taking blood away from the heart and one pumping blood toward the heart. The vessels that take blood away from the heart are called arteries and the vessels that return blood toward the heart are called veins. Veins have a very thin muscular layer that surrounds the vessel. Arteries have a much thicker layer because of their function. Arteries have to pump the blood out to the extremities in muscular contractions and wave-like motions.

The muscles in the circulatory system's vessels are controlled through a delicate balance of blood pressure. These muscles can shunt blood either to deep internal organs for digestion and regenerating capacities, or to extremities and skeletal muscle to respond to a parasympathetic signal. The body controls smooth muscle automatically through the autonomic nervous system.

Cardiac Muscle

Cardiac muscle is an involuntary striated muscle. Cardiac muscles contract automatically to squeeze the walls of the heart inward. The heart beats nonstop about 100,000 times each day. It can do this because of the cardiac muscles. Most of the heart is made of cardiac muscle which is unique to the heart .The heart has its own electrical pacemaking system that stimulates the cardiac muscle in a specific rhythm.

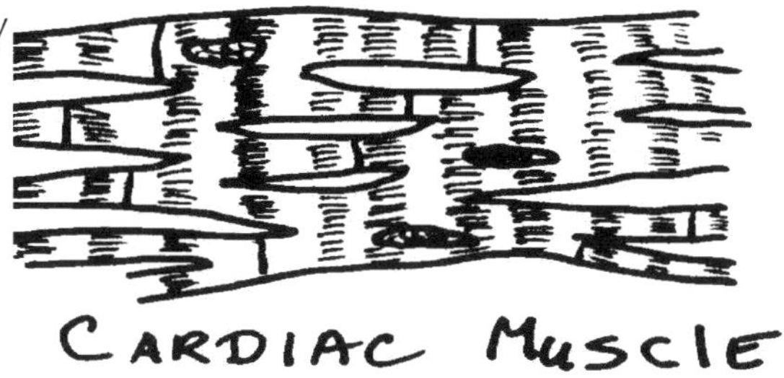

Cardiac Muscle

Visceral smooth muscle and cardiac muscles are controlled under an involuntary or autonomic nervous system. This autonomic nervous system, although not under our direct control, may be encompassed through expanded consciousness. Regular practice of long, deep breathing oxygenates the body and through relaxation techniques, there is an opportunity to learn voluntary control over the autonomic nervous system.

Skeletal Muscle

Skeletal muscle is a voluntary striated muscle. Skeletal muscles move and maintain the posture of the body. They make up fifty percent of your body weight. There are 640 individually named skeletal muscles. A skeletal muscle links two bones across its connecting joint. When these muscles contract or shorten, your bone moves. Muscles are arranged in layers over the bones. Those nearest to the skin are called superficial muscles. Those closest to the inside of the body are called deep muscles.

Voluntary muscles such as the skeletal muscles are directly under control from the somatic (body) nervous system and they are moved through conscious control.

The term "posture" means simply position or alignment of body parts. "Good posture" means many things. It means body alignment that most favors function. It means position that requires the least muscular work to maintain, which puts the least strain on muscles, ligaments, and bones. It means keeping the body's center of gravity over its base. Good posture in standing position, for example, means feet placed parallel firmly on the ground about 12 inches apart, knees bent slightly, the tailbone pushed towards the floor while lifting the pelvic area towards the navel, arms hanging near the body while pressing the shoulder blades backwards, head and heart held high and chin slightly in.

SKELETAL MUSCLE

Since gravity pulls on the various parts of the body at all times, and bones are too irregularly shaped to balance themselves on each other, the only way the body can be held upright is for muscles to exert a continual pull on bones in the opposite direction from gravity. Gravity pulls the lower jaw downward; muscles must pull upward on it. Muscles exert this pull against gravity by their ability to contract as well as sending and receiving messages from the nervous system, developing balance. When the body is in balance, with core strength and flexibility, and the mind is linked with the breath, the human form moves with grace and ease.

Sensors for the Motor System

To effectively control muscle the Central Nervous System needs information about the lengths of the muscle and the forces they are generating. The skeletal muscles are abundantly supplied with a variety of receptors. Two receptors are particularly important for motor control: the muscle spindles and Golgi tendon organs.

Muscle Spindles

Muscle spindles are elongated structures within the fleshy portions of muscle; they are parallel to the skeletal muscle fiber. The muscle spindle sends information to the nervous system about either the muscle length or rate of change of its length.

When the muscle is stretched, so is the muscle spindle. The muscle spindle senses the change in length and how fast it changes and sends this information to the Central Nervous System. This triggers the stretch reflex, which attempts to resist the change in muscle length by causing the stretched muscle to contract. The more sudden the change in muscle length, the stronger the muscle contractions will be. This basic function of the muscle spindle helps to maintain muscle tone and to protect the body from injury.

One of the reasons for holding a stretch for a prolonged period of time is that as you hold the muscle in a stretched position, the muscle spindle becomes accustomed to the new length and reduces its signaling. Gradually, you can train your stretch receptors to allow greater lengthening of the muscles.

Golgi Tendon

Golgi tendon organs lie in the collagen fibers of the tendon, in the area where the tendon and muscle fibers meet. When the muscles contract it places tension on the tendons where the Golgi tendon organ is located. The Golgi tendon organ is sensitive to the change in tension and the rate of change of the tension and transmits this information to the Central Nervous System. Golgi tendon organs help to prevent excessive stresses at joints by sending an impulse via afferent neurons to the CNS, where they synapse with motor neuron fibers of that same muscle. The efferent neurons instantly transmit an impulse to the muscle, causing it to relax, thereby preventing injury.

Posture and alignment

Postural muscles fix a joint and prevent movement while phasic muscles create movement. A postural muscle is one that is always working to maintain our posture. Postural muscles are stronger and deliver more work than phasic muscles. They have slower twitch muscle fibers, which can work longer.
Postural muscles are built for endurance and when they get tired they tighten up. Phasic muscles are more for movement and not much endurance. When phasic muscles are tired, they get stretched. Strengthen what you want to stretch, and stretch what you want to strengthen.

Deep Postural Muscles

These muscles are structured to be able to contract for long periods of time without fatiguing. Postural muscles are:

Adductor Longus
Adductor Magnus
Gastrocnemius
Iliopsoas
Latissimus Dorsi
Levator Scapulae
Lumbar Erector Spinae
Oblique Abdominals
Pectoralis Minor
Piriformis
Quadratus Lumborum
Rectus Femoris
Sacrospinalis
Semimembranosus
Semitendinosus
Soleus
Sternocleidomastoid
Tensor Fascia Lata
Tibialis Posterior
Upper Pectoralis Major
Upper Traps

Major Phasic Muscles

These muscles are designed to be able to respond and contract quickly. Examples of these muscles are:

Anterior Neck Flexors
Arm Extensors
Deltoids
Gluteus Maximus
Gluteus Medius
Gluteus Minimus
Hamstrings
Lower Pectorals
Lower Traps
Middle Traps
Peroneal Brevis
Peroneal Longus
Rectus Abdominis
Rhomboid Major
Rhomboid Minor
Scalenii
Serratus Anterior
Vastus Intermedius
Vastus Lateralis
Vastus Medialis

Muscle Names and Functions

A good way to start the study of muscles is to find out what its name means. Muscles are grouped according to function.

The following terms are used to designate muscles according to their main actions:	
Flexors:	Decrease the angle of a joint (between the anterior surfaces of the bones except in the knee and the toe joints)
Extensors:	Return the part from flexion to normal anatomical position; increase the angle of a joint
Abductors:	Move the bone away from midline
Adductors:	Move the part toward the midline
Rotators:	Cause a part to pivot on its axis
Levators:	Raise a part
Depressors:	Lower a part
Sphincters:	Reduce the size of an opening
Tensors:	Tense a part, that is, make it more rigid
Supinators:	Turn the hand palm upward
Pronators:	Turn the hand palm downward

Reasons for names: Muscle names seem more logical and therefore easier to learn when one can understand the reason for the names.

Each name describes one or more of the following features about the muscle:	
Its action:	flexor, extensor, adductor, etc.
Direction of its fibers:	as rectus or transversus
Its location:	as tibialis or femoris
Number of divisions composing a muscle:	as biceps, triceps, or quadriceps
Its shape:	as deltoid (triangular) or quadratus (square)
Its points of attachment:	as sternocleidomastoid

A Glimpse at the Relationship to Other Systems

Moving facilitates good brain function.
Movement increases blood circulation. The body adapts itself to the increased energy needs of the muscles by increasing the heart rate and therefore the amount of blood circulation that reaches the muscles, intensifying the breathing and oxygenating the blood.
Related organs and body systems benefit by relaxing reflexively.

Some Good Foods for This System

- Pineapple juice
- Yogurt and whey
- Walnuts and pistachios
- Mung beans and rice

How Yoga Helps

- Offers a balanced exchange of contraction and relaxing.
- Relaxes the muscles through stretching and motion.
- Muscle relaxation affects a specific area and a specific group of muscles.
- Muscle relaxation remains long after the yoga posture is complete. As a result, related organs and body systems are relaxed reflexively.
- A strong conservation and increase of energy is produced by the calming of body processes and of activity in the nerves. This energy is used for expanding the consciousness and for meditation.
- Movement increases blood circulation. The body adapts itself to the increased blood circulation that reaches the muscles, intensifying the breathing and thereby oxygenating the blood.
- Breathing powerfully while doing yoga is helpful for the muscles.
- Yoga keeps the joints flexible by cleaning deposits on the joints and avoiding new deposit build up.
- For an in depth look at musculoskeletal motion read the 13 Principles of Yoga-Motion chapter.

Precautions Regarding Yoga and the Muscular System

Be sure that proper warm-ups are used, if necessary, before proceeding to a kriya.
When teaching any pose that involves lifting the head, it is important that the neck lock is applied. Never put body weight on the neck. We do not do head stands for several reasons: The cervical spine has discs the size of a dime. Putting full body weight on that small of a surface area is not advised. The upper cervical spine is not stable under axial compression and head stand may cause it injury or even be fatal. There are small neurons in the brain that are not made for long periods of increased pressure. The most mobile region of the cervical spine is also the most vulnerable to injury, the upper cervical spine (C1 and C2).
Cobra Pose is a problem for many people. Stretch the front of the body so the student is not riding on the joints. Push the chest forward. Use a modified cobra pose when necessary.
Kundalini Yoga does not have a pose using headstands. Never put pressure on the cervical vertebrae. Shoulder stand is good; make sure the weight is balanced on the shoulders, not on the neck.

When teaching Plow Pose – go easy. It is better to increase the mid-back stretch before going fully into the pose.
Generally speaking, it is good to avoid bouncing stretches, which provoke a signal to the muscle to tighten up. Gradual, consistent stretching is to be encouraged.
It is important to synchronize the breath when stretching.
Encourage students not to be competitive, not to push too much; more progress is made when moving slowly and consistently.
Remember that regular and sufficient rest is just as essential to muscle function and development as regular and appropriate activity.

Yoga That Affects The Muscular System

Most kriyas in Kundalini Yoga will in some way affect the muscles and joints. Many Kundalini Yoga sets strengthen major muscle groups. A rather vigorous example is Flexibility and the Spine (p. 47-50, Sadhana Guidelines). It provides a good workout.

Common Muscles in Their Relationship to Yoga	
Biceps:	This is the only muscle that should be active during Life Nerve Stretch.
Triceps:	This muscle is strengthened during Triangle Pose pushups.
Deltoid:	The deltoid muscle lifts the arm up.
Latissimus dorsi:	This is a major muscle of extending the torso — Bow Pose.
Pectoralis major:	A major chest muscle used in push-ups.
Abdominus rectus:	The muscle used in "crunches" up to the angle of 20 degrees; after that, the iliopsoas is activated.
Iliopsoas:	The deep muscle of the trunk balanced during Stretch Pose; often involved in low back pain problems.
Quadriceps:	The muscles of the anterior thigh; to strengthen: Frogs.
Biceps femoris:	The muscle known as the "hamstring" muscle; antagonist to quadriceps.
Gastrocnemius:	The calf muscle; strengthened during toe raises.
Piriformis:	This muscle located across the buttocks is used to extend the leg. The sciatic nerve often (1/3) goes through the belly of this muscle. A major muscle involved in sciatica; stretched during Cat Stretch.
Sternocleidomastoid:	This muscle connects the sternum (breastbone) to the clavicle (collarbone) to the bone in back of the ear (mastoid); often involved in traumatic neck injury.

Some Conditions of the Muscular System

Sprain/Strain: When joints are moved beyond anatomical limits resulting in injury. Injuries develop inflammation and scar tissue.
Some uncomplicated back pain is due to, altered muscle imbalance and less than optimal spinal mechanics. Yoga can be very effective in developing muscular balance in the spine. When there is radiating pain, numbness, or lack of muscle control in the arms or legs, or difficulty controlling urination or defecation, it is evidence of a significant neurological problem indicating prompt attention before doing yoga.

Vocabulary

Fascicle: A bundle of muscle cells bound together by connective tissue.
Isometric contraction: A contraction in which the muscle does not shorten (the load is too heavy), it produces no change in the length of the muscle, but its internal tension increases.
Isotonic contraction: A contraction in which muscle tension remains constant and the muscle shortens.
Latent period: Period of time between neurological signal and the onset of muscle contraction.
Motor unit: A motor neuron and all the muscle cells it stimulates.
Muscle fiber: A single muscle cell.
Myosin: Contractile protein present in muscle cells.
Myofibril: A rod-like bundle of contractile filaments found in muscle cells; composed of individual sarcomeres.
Sarcomere: The smallest contractile unit of muscle; contains myofilaments composed mainly of the contractile proteins, actin, and myosin.
Sarcoplasmic reticulum: The specialized endoplasmic reticulum of muscle cells.
Somatic: Of the body framework or walls, as distinguished from the viscera or internal organs.
T tubules: Extensions of the muscle cell plasma membranes (sarcolemma) that protrude deeply into the muscle cell.
Tetanus: A smooth sustained muscle contraction resulting from high-frequency neurological stimulation.

Study Points

1. What are the three types of muscles?
2. What connects muscle to bone?
3. How do the muscles talk to the brain?

Chapter Eight

The Digestive System

The Digestive System is often described as a tube within a tube. The space, defined by the mouth on one end and the anus on the other, is technically defined as a space within the body yet outside the body. The nature of the substance inside the digestive tube and the substance around it is dramatic. You can eat contaminated foods, with bacteria such as staph and strep, or eat spoiled food, and your stomach may hurt but you won't die. Staph infections of the blood on the other side of the intestinal wall can kill you within a few days. A ruptured appendix can kill you in a few hours. The structural integrity of the membranes between the digestive system and the rest of the body is directly related to the function of this system. The digestive system is regulated by the parasympathetic nervous system. The digestive system has three major functions:

1. To absorb and digest foods
2. To secrete digestive juices
3. To move food through from the beginning to the end

The Manifesto of Digestion

- ALL food must leave the digestive system within 24 hours.
- A large percentage (40-60%) of your food should be raw.
- Eliminating waste & internal cleansing should happen naturally and with ease.
- Why are most yogis vegetarians? To be more alkaline than acidic, to facilitate the meditative process, to facilitate elimination (apańa), to access unused basic amino acids, and to be kind to animal life.
- Weight Management: Know the basic mathematics: Blood Sugar, Insulin, Exercise, and Leptin.

Basic Structure/Function

The function of the digestive system is to reduce large food particles to a size and chemical composition that can be absorbed by the small intestine and the blood, and thereby be utilized by the body cells for energy, growth, and tissue repair. Also considered part of the system is the process of elimination.

Yogis have long understood the importance of the area around the navel point. This is the center of an intelligence that has an influence on the entire human system. This is referred to as the "gut brain." In fact, many brain hormones are manufactured in the intestines.

The digestive system has autonomic, sympathetic, and parasympathetic innervation, including many important nerves like the vagus, splanchnic and celiac, as well as plexi specialized for various organs. Parasympathetic nerves stimulate secretion, peristalsis, relaxation, and promote the digestive process. Sympathetic nerves decrease secretion of digestive juices and slow peristalsis, sending blood and energy elsewhere in the body.

Organs in This System

Mouth

The mouth is the food-input orifice.

Teeth

The teeth grind, crush, cut and tear food, and mix it with saliva.

Salivary Glands

These glands add saliva to begin the breakdown of starches. There are three paired glands located in the mouth: parotid, submandibular, sublingual.

Esophagus

The esophagus brings food from the mouth, past the glottis (the flap that closes off the breathing tube), and into the stomach.

Stomach

The stomach mixes food (peristalsis) and breaks down carbohydrates, proteins, and fats. The stomach acts as both a processing area and a storage area, secreting acid and enzymes to begin digesting proteins while it churns the food inside it. The stimulus for the acidic environment is chemical, mental, and emotional. Over stimulation of the parasympathetic centers of the brain interrupts digestion.

Gallbladder

The gallbladder stores bile, which emulsifies fat. It is connected to the small intestine by the bile duct. Bile is made in the liver.

Liver

The liver is the center of chemical reactions in the body. Most of the complex blood chemistry happens here. It is a very large organ located above the stomach on the right side of the body. It is like a giant factory that processes all the food nutrients into usable sugar, fat, and protein. It eliminates 50% of the toxins ingested from food; the kidneys eliminate the other 50%. If the liver is overworked it becomes sluggish, which can result in fogginess of the brain and can be related to emotions of anger and depression. Unlike any other organ of the body, if part of the liver is removed, it will grow back to full size. This is another example of the AMAZING self-healing mechanisms of the human body, and an expression of the intelligence innate to each cell, tissue and organ system. The liver is often referred to as the father organ of the body and the spleen as the mother organ.

Laws of the Liver

- Do not eat large meals late at night. It is hard on the liver. It is best to finish eating for the day by 4 or 5 pm. If you eat late, eat light (like fruit).

- Yogi Tea: Drink yogi tea daily, it is a tonic. It does not replace water, in fact is a mild diuretic, so drink a glass of water for every glass of yogi tea you drink.
- Special foods for the liver: Fresh & raw foods, beets & carrots, Yogi Tea, steamed vegetables, watermelon, grapefruit, sauerkraut, yogurt (goat), and good fats!
- Good fats: Not trans-fatty acids, good omega 3 & 6 & 9.
- Anger is a form of stress that comes from frustration and a need to be right. Managing stress is essential for good liver function.
- Exercise helps the liver understand good energy management by releasing and creating glycogen. Glycogen is the reserve sugars stored in the liver. Accessing and replenishing this reserve aids in running, swimming & walking: glucose & glycogen cycles
- Avoid true toxins that the liver has to detoxify: pesticides, artificial sweeteners, MSG, unnecessary pharmaceuticals and alcohol.

Pancreas

Located under the stomach towards the left side of the body, this organ plays a major part in digestion. The pancreas secretes pancreatin, which begins the breakdown of proteins. It produces enzymes that are secreted in the duodenum. It also controls the blood sugar levels. Diabetes is caused by a lack of production of the hormone insulin, which allows blood glucose to be moved across the cell wall for energy.

Appendix

This organ is a small outpouching from the beginning of the large intestine (the ascending colon). The appendix contains lymph tissue that plays a role in the development of the immune system and the development of antigens.

Small Intestine

Its main functions are food digestion and the absorption of nutrients. The small intestine combines the acidic mix of the stomach with pancreatic juices, bile from the liver and gall bladder, and with intestinal secretions. It also absorbs the end products of digestion. The parts of the small intestine are the duodenum, jejunum, and ileum.

Large Intestine/Colon

The large intestine is made up of the ascending, transverse, descending, and sigmoid colons as well as the rectum. It absorbs water, salts, and products of bacterial activity (vitamin K, B vitamins, and amino acids) from the remaining digestive mix while storing it and forcing it out of the body.

Rectum

The rectum is the waiting room for exit of waste products.

Anus

This is the exit control muscle. Waste output.

A Glimpse at the Relationship to Other Systems

Many hormones are manufactured in two places in the body, the digestive system and the brain. **Fire is the dominant element utilized in digestion.**

Skin and Hair

- Don't put anything on your skin that you wouldn't eat. Your skin is porous and absorbs what you put on it into your bloodstream.
- Good Digestion = Good Skin
- A yogurt bath is good for the skin.
- Oil the skin with almond oil before bathing.
- Use soap sparingly.
- Wash the hair every 72 hours or more often if you have been sweaty from exercise or activity.
- Take a cold shower daily.
- Some sun is necessary, too much is very bad.
- Drink plenty of good water.
- Do not smoke.
- Eat good fresh food - fruits and vegetables.
- Negative Influences for This System
- A diet of processed food, or fried food
- Eating late at night
- A sedentary lifestyle
- Alcohol — severely weakens the liver
- Stress
- Inadequate water consumption

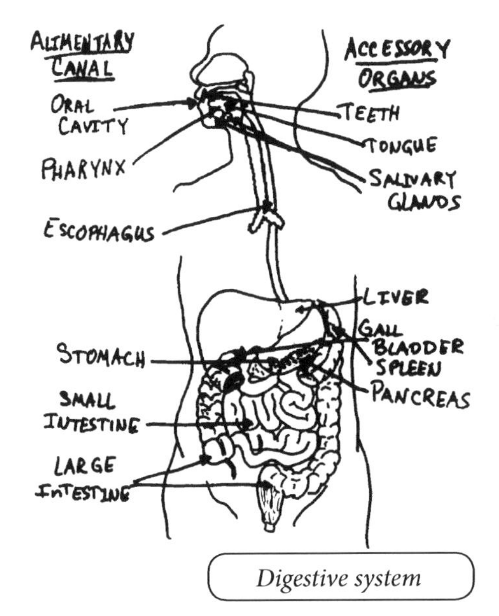

Digestive system

Some Good Foods for This System

- **For the stomach:** yogurt, wheat germ, brewer's yeast, peppermint tea, and ginger, fennel tea
- **For the liver:** beets, carrots, green foods
- **For the gallbladder:** olive oil, cold-pressed oil, whole grains, nuts, carrots, beets, pear juice, grapefruit juice, lemon juice, grape juice
- **For the small intestine:** wheat germ, green leafy vegetables, cabbage, oranges
- **For the large intestine:** lots of water, green leafy vegetables, yogurt (acidophilus)
- **For the kidneys:** lots of water, green vegetables, basmati rice, celery, cucumber juice, corn silk tea

How Yoga Helps

- Increases the fire energy at the navel point to aid digestion.
- Increases the secretion of gastric juices and digestive enzymes, and speeds up a sluggish metabolism.
- Increases the ability of the "gut brain" to self-regulate, changing the neurochemistry of the brain.
- Helps to nurture the environment in the intestines so that the beneficial bacteria may prosper and develop an evolving community.

- Increases the ability of the body to eliminate.
- Facilitates liver and kidney function.

Yoga That Affects The Digestive System

- Sat Kriya, Bow Pose, Camel Pose; sideways twisting exercises
- Elimination (Apana) Exercises (p. 55-56, Sadhana Guidelines)
- The Nabhi Kriya sets; for example: Nabhi Kriya for Digestion (p. 45, The Aquarian Teacher Level 1 Yoga Manual)
- The Art & Equilibrium of the Stomach (p. 39, Owner's Manual for the Human Body)
- Strengthening the Stomach (p. 40, Owner's Manual for the Human Body)
- Exercise Set for the Liver, Colon and Stomach (p. 14-17, Kundalini Yoga for Youth & Joy)

Some Conditions of the Digestive System

Inflammatory Bowel Disease (IBD): IBD describes conditions with chronic or recurring immune response and inflammation of the gastrointestinal tract
Irritable Bowel Syndrome (IBS): IBS does not produce the destructive inflammation found in IBD. It is a functional gastrointestinal syndrome, meaning there is some type of disturbance in bowel function.
Celiac disease: An inflammatory bowel disease of the small intestine. It is an immune reaction to eating gluten, a protein found in all forms of wheat (including durum, semolina, spelt, kamut, einkorn and faro), and related grains; barley, rye, and triticale. This immune reaction causes damage to the small intestine and does not allow food to be properly absorbed.
Crohn's disease: A form of inflammatory bowel disease that commonly affects the last part of the small intestine and parts of the large intestine. However, it isn't limited to these areas, and can attack any part of the digestive tract. Crohn's disease causes inflammation that extends deep into the layers of the intestinal wall.
Ulcerative colitis: A form of inflammatory bowel disease of the large intestine, where the inner lining of the intestine becomes inflamed and develops ulcers. It is often the most severe in the rectal area, which can cause frequent diarrhea. Mucus and blood often appear in the stool if the lining of the colon is damaged.
Constipation: Full of feces.
Diarrhea: The frequent passing of loose or watery stools.
GERD: Stands for Gastroesophageal Reflux Disease, a disorder in which there is recurrent return of stomach contents back up into the esophagus, frequently causing heartburn, a symptom of irritation of the esophagus by stomach acid.
Hemorrhoids: Dilation of the blood vessels of the anus, due to pressure from chronic constipation or increased portal hypertension.
Hiatal hernia: When the stomach or parts of it are squeezed above the diaphragm.
Irritable bowel syndrome, or IBS: A disorder of the loser intestinal tract. It involves abdominal pain and abnormal bowel movements. Emotional stress often makes the symptoms worse.
Eating disorders: A group of conditions characterized by abnormal eating habits that may involve either insufficient or excessive food intake to the detriment of an individual's physical and emotional health. Anorexia nervosa (non-eating), bulimia nervosa (habitual vomiting), and binge eating are the most common. Approximately 5% of adolescent and adult women and 1% of men have an eating disorder.

Vocabulary

Bile: A yellow-green fluid made by the liver and stored in the gallbladder. It passes through the common bile duct into the duodenum where it helps to digest fat.

Carbohydrates: Mainly sugars and starches, together make up one of the three nutrients used as energy sources by the body.

Enzymes: Proteins that act as a catalyst in mediating and speeding a specific chemical reaction.

Fat: Together with proteins and carbohydrates, one of the three principal types of nutrients used as energy sources (calories) by the body. The energy produced by fats is 9 calories per gram. Proteins and carbohydrates each provide 4 calories per gram.

Palate: The roof of the mouth. The front part is bony (hard palate), and the back part is muscular (soft palate).

Peristalsis: The rippling motion of muscles in the digestive tract.

Pharynx: The hollow tube that starts behind the nose and ends at the top of the trachea and esophagus. It is about about 5 inches long.

Protein: A large molecule composed of one or more chains of amino acids in a specific order determined by the base sequence of nucleotides in the DNA coding for the protein.

Saliva: A watery secretion in the mouth produced by the salivary glands that aids in the digestion of food and moistens and cleanses the mouth.

Throat: The throat is the front portion of the neck beginning at the back of the mouth, consisting anatomically of the pharynx and larynx. The throat contains the trachea and a portion of the esophagus.

Tongue: A strong muscle anchored to the floor of the mouth.

Study Points

1. What structures of the Digestive System must food pass by to go through the digestive system?
1. Where are most foods absorbed?
1. What does the liver do?
1. Where is the pancreas and what is its function?
1. Where does the gallbladder get the bile?
1. What does bile do?
4. What foods are broken down by saliva?

Chapter Nine

The Urinary System

Basic Structure/Function

The Urinary System consists of two kidneys connected to a storage area (the bladder) by long tubes (ureters). A third tube, the urethra, leads from the bladder to outside of the body.

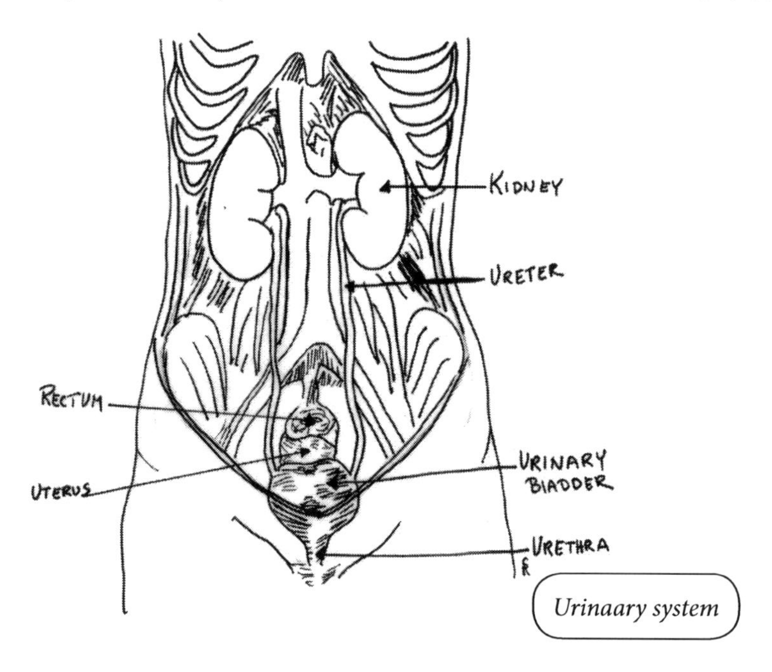

Urinaary system

Organs of This System

Kidneys

The kidneys are a silent, but a most important servant of the body. The major function of the kidney is to filter blood and to remove the fluid wastes of the body. The fluid is filtered through a system regulated by a balance of blood pressure, blood sugar, and water balance. Both a beautifier and a retainer of life, the kidneys are organs with which you cannot play games. Everyday the kidneys filter over 1700 liters of blood into about 1 liter of highly concentrated fluid called urine.

The kidneys are made of over 2,400,000 nephrons, each of which is capable of creating urine. The nephron is made of a complex system of membranes and tubes creating a variety of pressures and filters.

The kidney is innervated(connected to the nervous system)by the renal plexus, which comes from the vagus nerve and thoracic splanchnic nerves.

Sympathetic fibers stimulate hormone secretion, decrease kidney filtration rate, and facilitate reabsorption and contraction of the ureters.
Parasympathetic fibers cause relaxation of smooth muscle in the kidney itself as well as in the ureters.
The kidneys help maintain proper blood sugar, osmolarity and viscosity. The kidneys play an important role in the immune system and overall vitality. Heavy food causes the first injurious effects for the kidneys. It has been noticed that the people who eat meat, eggs, and fish suffer more with kidney disease than the vegetarian population.
Kidney health also affects the personality. Laziness, lack of initiative, lack of energy, irritability, a propensity to complain, and cowardly behavior are mainly attributed to poorly functioning kidneys. It is wisest to take care of the healthy kidneys instead of waiting for them to become unhealthy. The only way to maintain a healthy kidney and avoid illness is through healthy eating and drinking habits. Urine color is a good indication as to how the kidneys are functioning. The urine should be clear or straw in color.
There is less chance for a person to develop a kidney stone if they are drinking one half to one gallon of water each day; this is an absolute must for people who sit in chairs most of the time or for laborers who sweat a lot. The liver is also affected by kidney problems. Therefore good kidney health keeps the liver functioning properly.
Kidney health is easy to maintain, yet difficult to remember. Before retiring at night, drink a glass of water to keep enough water in the system for the night. In the morning, after clearing the monkey cavities, drink two pints of water to completely flush the kidneys.
Whenever there is trouble with the kidneys, first drink more water. Also, fresh homemade yogurt water (whey), watermelon, celery juice, cucumber juice watermelon juice, light food, and/or green food may help.

Bladder

The bladder collects urine. It can hold about two liters of liquid. When the energy of the bladder is disturbed, it can lead to tension in the back, the neck, the buttocks, the back of the thighs, and the outside of the feet.
Sympathetic nerves from the lumbar spine and parasympathetic nerves from the sacral plexus innervate the bladder. The sympathetic system relaxes the detrusor muscle of the bladder and contracts the internal sphincter.
Parasympathetic nerves contract the detrusor muscle and empty the bladder.

Ureter

The Ureter is a tube which transports the urine from the kidneys to the urinary bladder.

Urethra

The urethra is a tube that drains urine from the bladder out of the body. In females, the urethra is about 1½ inches long, ending at the vulva. In males it is about 8 inches long, ending at the tip of the penis. The urethra is a tube that carries urine (in females and males) and semen (in males) from inside the body to the outside

A Glimpse at the Relationship to Other Systems

The urinary and reproductive systems are often studied together because the systems are so intertwined anatomically. They share some common pathways neurologically and anatomically.

The kidneys filter out toxins that accumulate in the circulatory system(blood).
The kidneys respond both from neurologic and hormonal messages.
The kidneys control the water balance in the body. If they absorb more water back into the body, the blood volume goes up and so the blood pressure goes up. If they eliminate more water, then the volume goes down and the blood pressure with it. Because of this, unhealthy kidneys will put a great deal of stress on all other organs, and the heart in particular.
The adrenals sit right on top of the kidneys and communicate primarily hormonally.
The kidneys are regulated by vasopressin, a hormone synthesized in the hypothalamus and stored in vessels in the posterior pituitary.

Negative Influences for This System

The kidneys, along with the liver, are responsible for eliminating toxins from the body. This is why many types of drugs commonly damage them. There is evidence that common drugs such as Tylenol, ibuprofen, and aspirin can cause kidney failure even in moderate doses. Two common insults to the kidneys are the failure to drink enough water and long-term use of coffee and caffeine.

Some Good Foods for This System

- Kidneys
- Good, clean water, and lots of it
- All green vegetables
- Celery juice
- Cucumber juice
- Cornsilk tea
- Basmati rice
- Bladder
- Eggplant
- Okra
- Green leafy vegetables
- Carrots
- Wheat germ
- Pumpkin seeds
- Olive

How Yoga Helps

The kidneys sit right under the diaphragm. Improving flexibility in the spine and diaphragm opens up circulation and nerve impulses. Exercises like spinal flexes increase signals in the area, shifting circulation to the region.
Many elements of a yogic diet help to vitalize the kidneys and to eliminate additional stress.

Yoga That Affects The Urinary System

- Toning the Kidneys (p. 36, Owner's Manual for the Human Body)
- Exercise Set for the Kidneys (p. 22-23, Kundalini Yoga for Youth & Joy)
- Camel Ride
- Cat-Cow

Some Conditions of the Urinary System

Bladder and kidney infections: Bacterial infections can travel up the urethra (female more common) and cause a bladder infection. If left untreated, infection can move from the bladder to the kidney.
Diabetes: Excess serum sugar levels cause damage to the kidney tissues, causing leakage of proteins and sugars into the urine.
Drug-induced nephritis: Tissue damage from overexposure to anti-inflammatories and other

pharmaceuticals that damage the kidney tissues, such as: sulfonamides and some synthetic anti-biotics and penicillin.

Bladder Health

The backpressure from stopping the stream while urinating will give the bladder a second rinsing out. The stopping and starting of the flow of urine helps tone the urinary tract and aids in flushing compounds that may be harmful to the prostate.

An easy exercise is to drink a lot of water and pump your belly. Lying down flat on the floor, raise the legs up to 60 degrees and scissor them while breathing long and deep.

Suggestions for Bladder Infections in Females

- Avoid excess sugars.
- Wear cotton underwear.
- Drink more water.
- Drink pure cranberry juice (not cranberry cocktail).
- When nature calls - answer the call and go.

Vocabulary

Nephron: The structural and functional unit of the kidney. It consists of the glomerulus and renal tubule.

Glomerulus: A cluster of capillaries forming part of the nephron; forms filtrate.

Juxtaglomerular apparatus: The cells located close to the glomerulus, which play a role in blood pressure regulation by releasing the enzyme renin.

Glomerular Filtration Rate (GFR): The rate of filtrate formation by the kidneys.

Aldosterone: The hormone produced by the adrenal cortex that regulates sodium ion reabsorption.

Antidiuretic hormone: The hormone produced by the hypothalamus and released by the posterior pituitary. This is the hormone that stimulates the kidneys to reabsorb more water, reducing urine volume.

Micturition: Urination, or voiding; emptying the bladder.

Study Points

1. What gland sits on top of the kidneys?
2. What is the major function of the kidneys?
3. What are the benefits of interrupting the urine stream during urination?

CHAPTER TEN

THE REPRODUCTIVE SYSTEM

MALE REPRODUCTIVE SYSTEM

Basic Structure/Function

The functions of the Reproductive System are notably different from the functions of any other system of the body. Their proper functioning ensures survival, not of the individual as other organ systems do, but of the species.

The male reproductive system consists of organs that produce, transfer, and ultimately, introduce mature sperm into the female reproductive tract where fertilization can occur. Also, the testes secrete androgens, or male sex hormones, notably testosterone.

The reproductive role of the female is much more complex. The female system must not only produce gametes (eggs), but the body must also prepare to nurture a developing embryo for nine months.

ORGANS IN THIS SYSTEM

Testes

Testes secrete testosterone, the primary male sex hormone, and make sperm in a process called spermatogenesis.

Epididymis

This organ stores the mature sperm.

Vas Deferens

This is the conduit for sperm during sexual activity.

Seminal Vesicle and Prostate Gland

They produce the plasma, sugar, and prostaglandin of the sperm, which is essential for reproductive success.

Ejaculatory Duct

The ejaculatory duct allows the release of sperm. It delivers sperm and eliminates urine through its tube. It also has inflatable tissue that acts as a major organ of delivery of sperm into the vagina, near the cervix (neck) of the uterus.

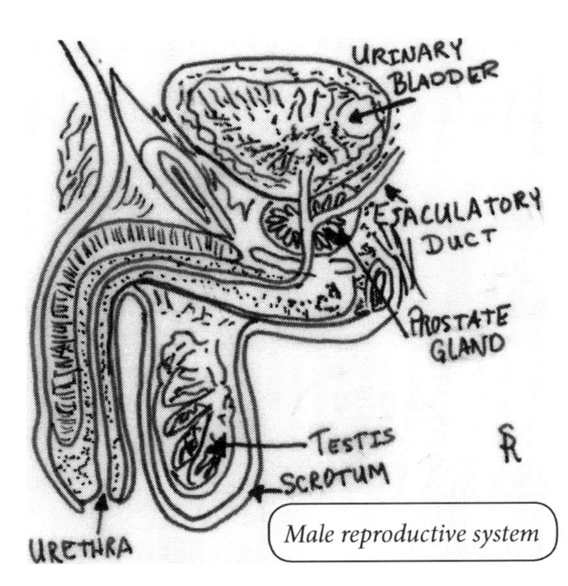

Male reproductive system

Scrotum

The scrotum contains testes and the vas deferens.
It controls temperature of the testicles for optimal production and is controlled by cremasteric muscle. The habit of wearing kacheras or boxer shorts during cold showers lessens the shock on this temperature-sensitive system.
In healthy males the ejaculate contains about 100 million sperm per milliliter. There are several types of sperm cells and not all target the egg. Some serve to block other suitors. Some help create the proper chemical environment for fertilization, and about 20% are deformed with no tails.
Sympathetic nerves from the lumbar and thoracic spine and parasympathetic nerves from the sacral plexus control the male reproductive organs. The pelvic splanchnic nerves and the pudendal nerve innervate this highly sensitive region.
Erection is brought about by a parasympathetic response stimulated by tactile stimulation of the erogenous zones or by the messages of sight, smell, sound, and thought from the brain.
Ejaculation is controlled by the sympathetic nervous system, from the lumbar spine.

Some Good Foods for Male Sexual Vitality

- Garlic
- Onions
- Ginger
- Black sesame seeds
- Black garbanzo beans
- Fruits that start with the letter "P"
- Figs and saffron figs
- Food cooked in ghee (butter oil)
- Refer to: Man to Man series by Yogi Bhajan

Good Sexual Habits for Men

- Refrain from eating food within 2 to 4 hours before intercourse.
- Urinate with intermittent stream within 15 minutes of ejaculation.
- Take a shower before intercourse begins, use almond oil on the upper legs and genitals.
- Masturbation (for men) weakens the nervous system, causes excessive brain pressure, and is in general a waste of vital energy reserves.
- Tight underwear causes excessive heat in the testicles. Kacheras/boxer shorts help maintain the sperm regulatory system.
- The aura of the man can either give energy or take energy according to the strength, clarity, and purity of the partner's electromagnetic field. Sexual play is said to start at the female pituitary gland.

Yoga That Affects The Male Reproductive System

- Chair Pose
- Frog Pose
- Life Nerve Stretch
- Archer Pose
- Sat Kriya

Female Reproductive System

Basic Structure/Function

The primary focus of the female reproductive system is to contain her eggs and to provide a nurturing environment for the development of an embryo (future baby). A woman is born complete, with ALL her eggs. The internal nature of the female reproductive system is reflected in the vast internal universe and psyche of the woman. The nature of woman is nature itself; there is no distance between woman, nature, God, and the universe.
The human female reproductive cycle is initiated and maintained by a variety of hormones. The experience of the human reproductive cycle is an ever-fluctuating combination of emotion, passion, reflection, and flow. The female psyche revolves around ovulation and is in rhythm with the phases of the moon. The female cycle reflects in the consciousness of the woman through the moon centers. Throughout the month, different moon centers are in effect. During foreplay, the moon centers should all be stimulated, to effect a complete and healthy tiding during sexual pleasure.

Organs in this System

Ovary

Ovaries release the mature egg into the fallopian tubes.

Uterine (Fallopian) Tube

This organ delivers eggs to the uterus.

Uterus

The uterus is the place of egg gestation (nine months).

Vagina

The vagina is the birth canal. Bartholin's glands and Skeen's glands are glands inside the vagina that secrete lubricating secretions.

External Genital Structures

Labia Major

The outside lips of the genital structure.

Labia Minora

The inner lips of the genital structure.

Clitoris

The clitoris is the female analog of the penis; it has erectile tissue and is approximately three inches long. It has 3x as many nerves as the penis.

Paraurethral Glands and Vestibular Glands

These are the glands of lubrication.

G-spot

The G-spot is the interior root of the clitoris

Vaginal Orifice

This is the outside opening of the birth canal.

Perineum

The perineum is the space between the anus and the lower vagina. In oriental medicine, the two major meridians gap at two places – at the perineum and at the mouth. These are the gateways to the two caves of creation (goofas). Parasympathetic and sympathetic nerves innervate the female reproductive system. Sympathetic nerves stimulate uterine contraction, and supply vaginal arteries and glands as well as erectile tissues. The parasympathetic nerves supply the layers of the vagina and urethra as well as the clitoris and other glands.

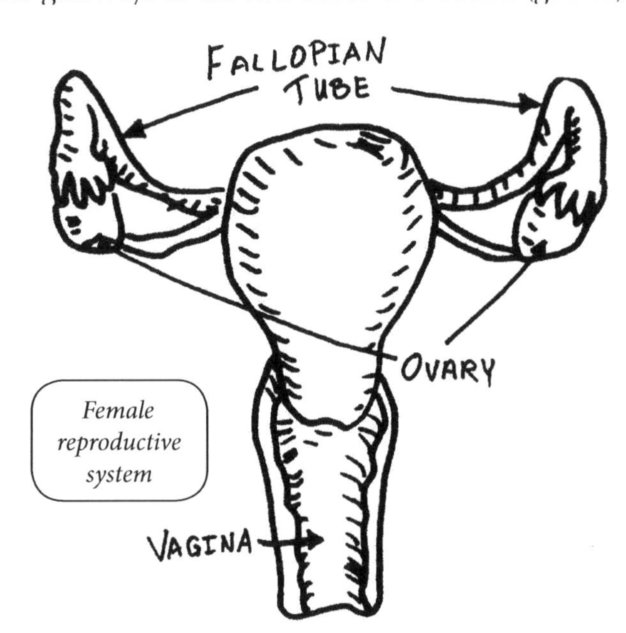

Female reproductive system

Precautions Regarding Yoga and Women During Menstruation

Inverted exercises such as shoulder stand and plow pose should be avoided during the menstrual flow. The inversion can cause undue stress on the flow of apana. Breath of Fire should be avoided during menses – light Breath of Fire may be done on lighter days.

Any abdominal exercise such as Stretch Pose, Maha Mudra, leg lifts, and bicycles, should be avoided during menstrual flow.
Women who relax and meditate during this time have noticed they have less cramping and discomfort.
Healthy bowel elimination is essential for a healthy female reproductive system.
Stagnated and strained circulation prevents the balanced inflow and outflow of hormones and their by-products.

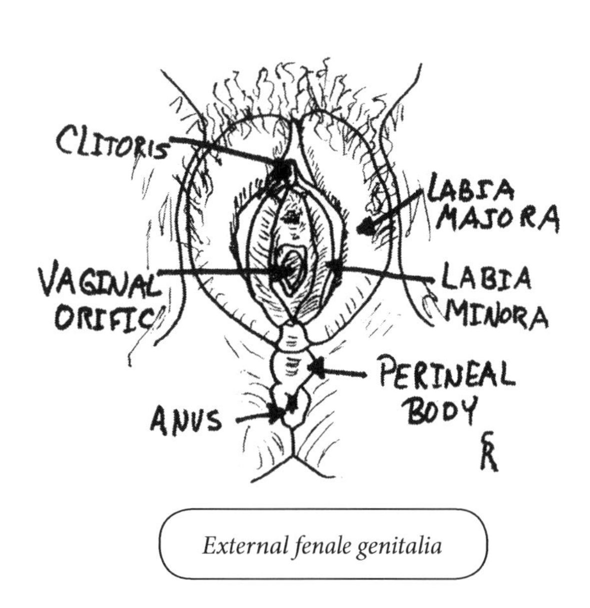

External fenale genitalia

Some Good Foods for Female Sexual Vitality

- Green chilies
- Rice bran syrup
- Citrus, plums, peaches, papayas, raisins, and dates
- Watermelon, bananas, mangos, guava, seeds, beets, beet greens, all other fresh vegetables
- Turmeric, yogurt, and wheat berries also help keep the system clean

During menstrual flow:

- Almonds cooked in almond oil with skin on (otherwise the skin should be removed); serve with honey

Yoga That Affects The Female Reproductive System

- Magnificent Seven Kundalini Yoga Exercises for Women (p. 270, Kundalini Yoga: The Flow of Eternal Power)
- Kundalini Yoga Kriyas and Meditations for Women (I Am a Woman:
- Essential Kriyas for Women in the Aquarian Age)
- "Grace of God" meditation (p. 85, A Woman's Book of Yoga)
- Meditation on the Divine Mother, Adi Shakti (p. 136 - 138, A Woman's Book of Meditation)
- Meditation to Balance the Moon Centers*
- Women can benefit from these meditations by doing them for at least 40 days.

Special Meditation for Balancing the Moon Centers

This meditation is the highest meditation for a woman. One can break any habit with it. It balances the moon centers, the outer moon effect, the inner moon cycle, and one's own zodiac moon effect. It also balances the glandular system.
Directions: Lie on the stomach, place the chin on the floor and keep the head straight. Arms should be along the sides of the body with the palms facing up.
The Panj Shabad is the mantra for this meditation.
- Saa - Infinity, cosmos, beginning
- Taa - Life
- Naa - Death
- Maa -Rebirth

The mantra is chanted silently with the eyes focused on the third eye (brow point). Meditate on the sound current coming in through the crown (the tenth gate) and flowing out through the third eye point, creating an "L" with each sound.
As you mentally vibrate Saa press the index finger to the thumb with a firm pressure, on Taa press the middle finger to the thumb, on Naa press the ring finger to the thumb, and on Maa press the little finger to the thumb. The breath will regulate itself. Continue in this manner for 3 – 11 minutes.

Pregnancy and Kundalini Yoga

This is time to practice yoga and meditation on a regular schedule and to proceed with caution. The basic guideline is to use common sense: Crow Pose, life nerve stretch, butterflies, and gentle pranayama. Meditation has a powerful effect on the child. Meditation before the

120th day after conception can affect the soul's progressing. See Conscious Pregnancy: The Gift of Giving Life for more information on childbearing and development. Also, see, I Am A Woman Book and Yoga Manual and the Khalsa Women's Training Course Manuals from Kundalini Research Institute (KRI) for more extensive teachings on the art of being a woman.

The Moon Centers

There are eleven sites of the moon in every woman. These eleven places are where the moon moves around her chin spot. The cycle runs complete every 28 days; the moon moves to a different center every 2 days. In the 28 days of her menstrual cycle, her moon can be found in eleven areas of the body, called the eleven limbs of a woman. The essence of a woman is moon in quality. The way in which her moon moves will vary in every woman and her mood will change as her moon changes.

The pattern of the moon stays the same in each individual but varies from woman to woman. Here is how every woman can know where her moon is: draw a picture of a woman, marking the eleven moon spots. Then make a chart of your basic or elementary behavior. In this way you can find your pattern and understand your behavior.

Women's Moon Centers

Moon Center #	Part of the Body	How You are Affected
1)	Hairline	This is the most important in terms of sensitivity. Nothing can move you an inch; so, in reality, you are real when your moon is in the hairline, your arc-line, your halo
2)	Pink of the cheeks	The pink of the cheeks is a very dangerous spot; it is when you are absolutely almost out of control
3)	Lips	
4)	Ear lobes	Called the sunspot; you will always discuss values
5)	Back of the neck	Wanting to communicate on a very romantic level; this is such a "foolish" time that one flower or little gesture can totally soften you
6)	Breasts	Compassionate and giving to the extent of foolishness
7)	Belly button area and the corresponding area on the back of the spine	This is when you are most insecure

8)	Inner thigh	You are very confirmative; you want to confirm everything
9)	Eyebrows	Most imaginative, illusionary, and tend to build sandcastles. When the moon is in the eyebrows, all a man has to do is trace your brow and you will fall right over
10)	Clitoris	Eager to socialize, talk, meet. Basically you are very external; your extra-curricular activity is very charming
11)	Membrane in the vagina	Your mood is to socialize

In the male, the moon center changes are controlled by hair. The male moon center is in the chin; it has hair on it, therefore, it is steady. The female moon center, also is the chin, but it does not have hair; therefore, the female fluctuates more than the male. It is like the waxing and waning of the moon.

Vocabulary

Bartholin's gland: It is a tiny gland inside each of the labia near the opening of the female vagina, one gland on each side. These glands secrete mucus to moisten the labial opening of the vagina.

Bulbourethral glands: They are located behind and to the side of the male urethra. They discharge a component of seminal fluid into the urethra. There are two bulbourethral glands, one on each side. They are the counterparts of Bartholin's glands in the female.

Gonad: Primary reproductive organ; i.e., the testes of the male or the ovary of the female.

Gamete: Sex cell; a sperm or ovum.

Seminiferous tubule: Site of sperm formation in the testes.

Semen: Fluid mixture containing sperm and secretions of the male accessory reproductive glands.

Skene's gland: It is on each side of the female urethral opening. The Skene's gland is similar to the male prostate gland. The gland produces lubricating fluid during female ejaculation that has a similar consistency to male prostatic fluid.

Spermatogenesis: Process of sperm (male gamete) formation.

Oogenesis: The process of ovum (female gamete) formation.

Follicle: Ovarian structure consisting of a developing egg surrounded by one or more layers of follicle cells.

Ovulation: Ejection of an immature egg (oocyte) from the female ovary.

Puberty: Period of life when reproductive maturity is achieved.

Menopause: Period of life when ovulation and menstruation cease, owing to hormonal changes.

Study Points

1. What lifestyle habits promote sexual vitality?
2. Which division of the nervous system is used during sexual intercourse: Sympathetic or Parasympathetic?
3. What is the function of the kidneys, bladder, prostate, Skene's and Bartholin's glands, ureter and urethra?
4. How does yoga affect sexual vitality?

Chapter Eleven

The Immune System

The Immune System protects the body from pathogens and tumor cells.

Basic Structure/Function

Being sensitive enough to distinguish between good bacteria and lethal pathogens is not a simple task. Some viruses and parasitic worms use special tricks to become stealth invaders. They can even adapt and evolve to avoid detection. The immune response begins at the blood cell level. Every cell has an immune response, but the white blood cells in the blood are the first line of defense. There are specific white blood cells that attack bacteria, viruses, allergic sensitivities and other proteins detected that are not self.

The immune system adapts over time and has a memory to previous assaults. This is called acquired immunity. Infants have an immature Immune System when they are born and acquire their individual immune profile as they experience immune challenges.

The white blood cells work with a web of vessels called the lymphatic system. The nervous system has a profound effect on the Immune System. Your mental state also has an effect on your immune fitness. Stress of any type is detrimental to the immune system. Getting adequate sleep (6-8 hours) is optimal for an optimal immune function.

Organs in This System

Surface membranes

Skin is actually one of the most important organs of the Immune System. Skin acts as a barrier to keep potential invaders from entering the body. Acids found in sweat help by making the skin inhospitable to bacteria. The bacteria that naturally live on your skin can also be useful. They produce toxins to keep competing bacteria from entering their territory.

Bone marrow

All the cells of the Immune System are formed in the bone marrow by a process called hematopoiesis. This process involves differentiation of bone-marrow derived stem cells either into mature cells of the Immune System or precursor of cells which move out of the bone marrow and continue their maturation elsewhere. The bone marrow is responsible for the production of important Immune System cells like granulocytes, B cells, natural killer cells and immature thymocytes. It also produces red blood cells and platelets.

White Blood Cells

White blood cells, also known as leukocytes, are the front line of the immune system. They defend the body against infectious organisms and foreign substances. Each type of white blood cell has a specific role or immune function that it's destined to carry out.

White blood cells (WBCs) are produced in the bone marrow. When the red blood cells are removed from blood, a watery plasma or serum remains. White blood cells are suspended in this watery liquid.

The white blood cell count, which is included in the complete blood count (CBC), is a measurement of the number of WBCs present in one milliliter of blood. White blood cells proliferate, increasing dramatically, in response to bacterial infection. An increased count, which is called leukocytosis, can be seen in infection, stress, and in various blood disorders and malignancies including leukemia.

Thymus gland

The thymus gland is a small, butterfly-shaped organ which lies between your breastplate and your heart and is most active, producing scores of lymphocytes during the childhood days. It is fully developed at birth and grows until puberty, after which it becomes fatty and shrinks to about 15% of its maximum size.

During its most active time, the thymus is responsible for directing the maturation of immature thymocytes into T cells. T cells are like the managers of the Immune System, instructing other cells how to react to foreign substances.

During this maturation in the thymus, T cells learn to differentiate between "self" (the body's own cells) and "nonself" (foreign objects, organisms, or diseased cells). If a T cell thinks a self cell is foreign, it is destroyed as it could cause the effects of an autoimmune disease, if allowed to leave the thymus.

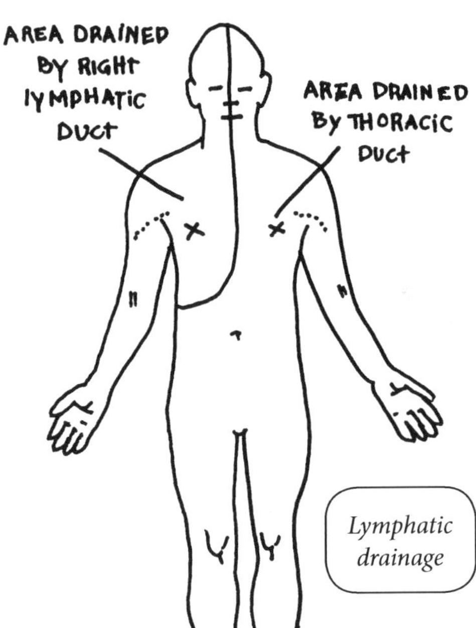

Lymphatic drainage

Spleen

The spleen is a spongy organ about the size of your fist and is located on your left side above your abdomen and underneath your lower ribs. It is composed of T cells, B cells, natural killer cells, macrophages, dendritic cells and red blood cells. The spleen has two main functions: to filter blood for foreign particles and dying red blood cells and to coordinate the immune response.

It acts as an immunologic filter of the blood and entraps foreign materials, that is antigens from the bloodstream passing through the spleen. When the macrophages and dendritic cells bring antigens to the spleen via the bloodstream, the B cells in the spleen get activated and produce large levels of antibodies. Thus, the spleen can also be known as the immunologic conference center. The spleen also forms the site of old red blood cells destruction. The spleen is often referred to as the mother organ and the liver as the father organ.

Lymphatic System

The lymphatic system is an extensive drainage network that helps keep bodily fluid levels in balance and defends the body against infections. It is made up of a network of lymphatic vessels that carry lymph throughout the body.

One of the lymphatic system's major jobs is to collect extra lymph fluid from body tissues and return it to the blood. This process is crucial because water, proteins, and other substances are continuously leaking out of tiny blood capillaries into the surrounding body tissues. If the lymphatic system didn't drain the excess fluid from the tissues, the lymph fluid would build up in the body's tissues, and they would swell.

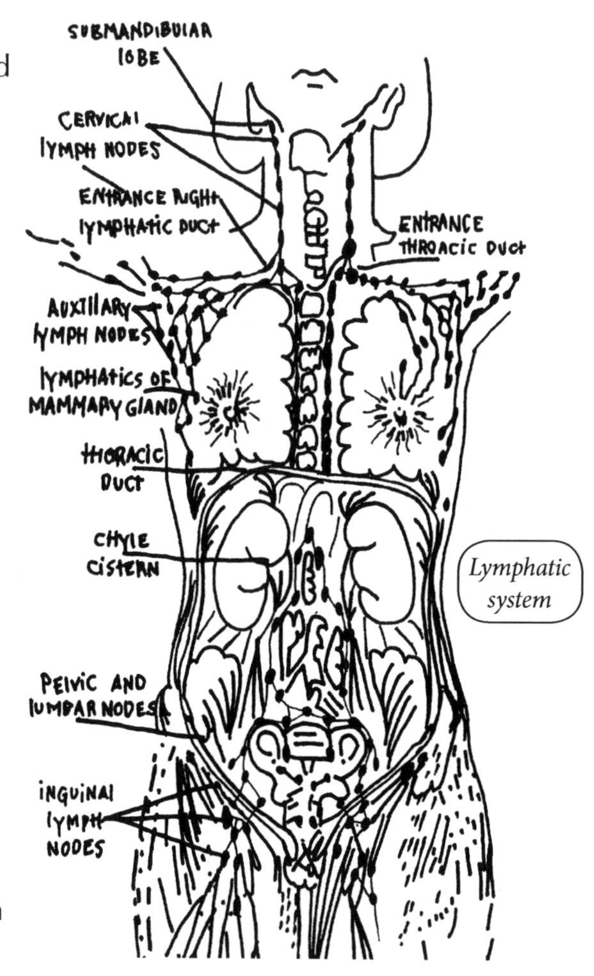

Lymphatic system

Lymph

The lymphatic system is made of a fine web of vessels, through which lymph flows. These are thin hair-like tubules that converge at nodules called lymphatic ganglia. These ganglia filter the lymphatic fluid, which eliminates bacteria and malignant cells, while producing white blood cells and processing waste. Approximately 10% of the fluid that leaves the arterial capillaries and penetrates the lymphatic vessel returns to the blood through the lymphatic system. The lymphatic system helps maintain proper osmotic and hydrostatic pressures in the circulatory system. The lymphatic system is controlled by parasympathetic nerves from cranial nerve X, the vagus nerve, as well as through sympathetic innervation from the brainstem.

Lymph-Vascular Division

The lymph-vascular division collects and returns tissue fluids to the vascular system. It starts out in the tissues as lymphatic capillaries, which begin with lymph nodes and form extensive networks. They are similar to blood capillaries in that they consist of one layer of endothelial cells, but they differ by being thinner and by not connecting with arteries. Lymphatic capillaries in the villi of the wall of the small intestine pick up fatty acids and glycerol from the digested food. These lymphatic capillaries are called lacteals, because of the milky appearance of their contents.

Lymphatic Vessels

Lymphatic vessels, formed where lymphatic capillaries come together, follow the veins in their distribution and may be divided, like veins, into superficial and deep groups. Lymphatic vessels, again like veins, are similar in the tissue makeup of their walls. Lymphatic vessels, however, are mostly thinner-walled and have more numerous valves, giving them a beaded appearance.

The lymphatic vessels from the right side of the head, the neck and the thorax, and from the

upper right limb join to form a common vessel, the right lymphatic duct. This duct empties into the junction of the right subclavian and the right internal jugular veins, thus mixing the lymph with the venous blood. The thoracic duct, where the lymphatic vessels from the remainder of the body lead into, originates in the abdomen as a chamber called the chyle cistern.
From there, it runs upward near the mid-dorsal line, then passes through the diaphragm with the aorta, and continues through the thorax joining the junction of the subclavian and the left internal jugular veins.

Lymphatic Nodes

The lymphatic nodes are found in the pathways of lymphatic vessels. Lymphatic nodes are masses of lymphoid tissue that range from the size of a pinhead to about the size of a lima bean or grape. They are found in groups of as few as two to as many as several hundred. Similar to the lymphatic vessels, some nodes are deep, whereas others are easily felt beneath the skin in the neck, the axilla, and the inguinal regions. These nodes serve to "filter" lymph, as well as to manufacture lymphocytes. Lymphocytes make up about 20 to 25% of the white blood cells.
The main clusters of lymph nodes are located in the head and face, neck, along the spine, thoracic region, armpits, lower extremities, pelvis, groin area, and abdomen.
As the lymph moves slowly through the node, the reticuloendothelial cells, by phagocytosis, "filter out" foreign particles, including bacteria. This prevents the entrance of foreign particles into the bloodstream. Infection can cause increased pressure in the lymph node causing tenderness and swelling. Contractions of body muscles pump lymph through the lymph ducts.

G.A.L.T. — Gut-Associated Lymphatic Tissue

This specialized tissue protects the digestive tracts and upper respiratory system. There are several isolated clusters of lymph nodules located in the throat and intestines. In the throat these tissues are called the tonsils. In the distal portion of the small intestine are located clusters of lymph tissue called Peyer's patches that regulate the growth of bacteria in the digestive tract.

Our Human Ecosystem

Bacteria

The organs of the immune system manufacture immune cells that provide defense against threat to the body by harmful bacteria and viruses. Bacteria are small organisms (animals) that find the human body a nice place to live. Not all bacteria are harmful to the human body. Most bacteria have a function in processes like digestion and skin maintenance. Bacteria like streptococcus and staphylococcus are common in our environment, and are consumed by the healthy immune system in quick order. Antibodies develop in the blood after exposure to specific bacteria, which alert the body to their presence upon exposure in the future. Antibiotic drugs have been developed to kill specific forms of bacteria when the body cannot.

Viruses

Viruses are very different from bacteria. Viruses are hijackers of cells. Viruses are mutant strands of DNA looking for a cell to express their genetic code. Viruses are usually recognized by the immune system; however, some are very tricky, like the AIDS virus, which changes before the body can respond. Viral infections create lots of "cell trash" causing muscles to feel sore and stiff, a common "flu" symptom. Lymphocytes, which mature in the thymus, or T-lymphocytes, fight viruses and have several functions: some are killer cells, some are suppressor cells, and some others have other specific functions. Antibiotic medicines do not affect viruses.

The Defenders

First Defense: Surface Membrane Barriers

Skin and all membranes secrete chemicals (sebum) toxic to bacteria. The pH of skin secretions inhibits bacterial growth. Bacteria (germs) are everywhere; this basic barrier prevents further advancement of the bacterial army. Harsh soaps and foaming agents (such as sodium laurel sulfate found in most mainstream shampoos, soaps and lotions) dissolve the protective layer of the skin and are not recommended.

The mucosal lining of the stomach secretes enzymes and a concentrated hydrochloric acid, both of which inhibit pathogens.

The oral cavity is cleansed by saliva and the external surface of the eye is cleansed by lacrimal fluid that we call tears. They contain lysozyme, an enzyme that destroys bacteria.

A sticky mucus is produced in the respiratory system and digestive system, which traps many microorganisms that may enter those passageways.

Nonspecific Chemical and Cellular Defense

- Phagocytes engulf and destroy bacteria (pathogens) that make it through the first defense.
- Natural killer cells attack cancerous and virus infected cells.
- Antimicrobial substances like interferons and complement factors mobilize the immune system.
- Inflammatory response aids in mending injured tissue.
- Fever is a systemic response to inhibit microbial multiplication and enhance tissue repair process.

Specific Body Defenses

There are several families of specific immune cells: the B-cells and three types of T-cells.

- B-cells produce antigens that are specific.
- Killer T-cells are cells that kill invading bacteria, cancer cells, parasites, and foreign cells by direct assault.
- Helper T-cells stimulate the production of other immune cells.
- Suppressor T-cells regulate the activity of T- and B-cells.
- Lymphokines are chemicals released by sensitized T-cells that facilitate communication between immune cells.

Immune Laws

- Self-esteem is essential for good immune health.
- Focusing on danger, fear or insecurity is injurious to the immune system.
- Don't worry, be happy.
- Avoid pesticides, get plenty of rest, exercise, and eat plenty of fruits and vegetables.

A Glimpse at the Relationship to Other Systems

The lymphatic system helps maintain the balance of fluid dynamics in the circulatory system. The lymphatic system joins the circulatory system near the shoulder.

There are many lymph vessels are in the intestines. The immune system creates a synergistic culture among bacteria and the body.

Mental or emotional stress can affect hormonal balance and thus affect the competence of the immune system. Under conditions of stress, the adrenal glands secrete adrenaline. Under the influence of adrenaline, the sympathetic nervous system is activated and redistributes blood away from functions like; repair, digestion, and metabolism.

The connection between stress and immune inhibition is often referred to as the "fight or flight syndrome" and has been firmly established since the 1950s. This was first described in Dr. Selye's book, "The Stress of Life." Psycho-neuroimmunology is continuing this fascinating exploration of the mind-body connection. More recently, Dr. Herbert Bension has done much research in this field.

A sense of worth and value creates a strong immune system by decreasing stress. Through knowing and respecting yourself, you are able to better experience the link with the infinite self, supreme consciousness, and destiny. These factors help establish a secure spiritual environment, which boosts the immune system.

Some Good Foods for This System

- Onion
- Ginger
- Garlic
- Fresh leafy vegetables
- Fresh fruit

How Yoga Can Help

In a controlled scientific study, done by the Kundalini Research Institute, blood samples were taken before and after a yoga set (the Kirtan Kriya), showing an increase in several pituitary hormones.

Yoga has a powerful cleansing effect and stimulates the blood circulation, the organs, and the immune system.

Sound and rhythm such as those used in yoga have a powerful effect on the immune system by stimulating the hypothalamus.

Kundalini Yoga has been shown to improve confidence, self-image, and self-worth.

Yoga That Affects The Immune System

Sets such as Immune Strengthening Workout (p. 41-44, Inner Workout) Set: Massage for the Lymphatic System (p. 8-9, Physical Wisdom); Exercise Set for the Lymph System (p. 18-20, Kundalini Yoga for Youth and Joy).

Pranayam

For the spleen: sitting in rock pose, hands on the thighs, hold out the exhale and pump the stomach

For the thymus: thumping on the chest

Kirtan Kriya

Some Conditions of the Immune System

Leukemia: Malignant white blood cells. A condition where the bone marrow is filled with non-functional cells.

Hodgkin's lymphoma & Non-Hodgkin's lymphoma: Cancer of the lymph system.

Lupus: Autoimmune disease. Antibodies against an individual's own DNA.
Human Immunodeficiency Virus (HIV): A mutating virus the human immune system can't seem to get rid of.
Acquired Immune Deficiency Syndrome (AIDS): HIV causes AIDS by damaging the immune system cells until the immune system loses its ability to defend against diseases caused by bacteria, viruses, and other microscopic organisms
Hypersensitivity: An intense immune system response to an allergen. The response often begins a cascade of immune reactions, creating an additional delayed hypersensitivity called cell-mediated hypersensitivity. Poison ivy is an example of a plant with a delayed hypersensitivity.
Autoimmune diseases: When the immune system develops antibodies toward its own tissues.
Anaphylaxis: When allergens cross-link with specific antibodies (IgE) beginning a cascade of events, creating either a local or systemic response of inflammation or spasm; this is usually initiated by a venomous bite or sting. Bites from spiders, snakes, scorpions, bees can be of this type and often require immediate medical attention. Systemic effects may include skin rashes like hives and bronchial constriction, excess mucus production, joint pain, and eye irritation.
The immune network and aging: The Immune System serves us very well until late in life. The ability to fight infection declines as people age.
Stress and depression: The mind has a powerful effect on the immune system. Tcell responsiveness is seriously depressed during times of stress or depression. The Nervous System is connected to the Immune System through the unique messages that direct the neurochemistry of the brain. Neuron cells produce receptors and neuropeptides, which have specific Immune System function and are a factor in mood and behavior. Hormones such as cortical steroids and epinephrine provide chemical links between the Nervous System and the Immune System.

Vocabulary

Chyle: An opaque, milky fluid absorbed by the villi of the small intestine, and carried by lacteals, a set of vessels similar to the lymphatics, to the beginning of the thoracic duct, where it is intermingled with the lymph and poured into the circulation through the same channels. It is the fat-containing lymph in the lymphatic of the intestine.

Ganglia: A cluster of nerve cell bodies occurring outside of the central nervous system.

Inguinal: Of the groin.

Lacteals: Lymphatic vessels of the small intestine.

Lymphatic: General term used to designate lymphatic system, including vessels and nodes.

Lymph: A transparent, colorless or slightly yellow fluid, which is conveyed by a set of vessels named lymphatic into the blood; the colorless, watery fluid exuded through capillaries to nourish body tissues; the fluid in the lymphatic vessels, similar to blood plasma.

Lymphocyte: One type of white blood cell.

Lymph node: Small glands that are scattered throughout the lymphatic system, and are found especially in the neck, throat (tonsils), armpits, groin, thighs, and body organs. Lymph nodes consist of networks of fibers and cells. The principal cells are called lymphocytes. Lymphatic nodes are masses of lymphoid tissue that serve to "filter" lymph, as well as to manufacture lymphocytes. Lymphocytes make up about 20 to 25% of the white blood cells.

Phagocytosis: A process by which a segment of the cell membrane forms a small pocket around a bit of fluid outside the cell, breaks off from the rest of the membrane, and moves into the cell. The process by which the white blood cells engulf and destroy bacteria.

Reticuloendothelial cells: Various types of connective tissue cells that carry on phagocytosis. They are widely scattered in the lining of the liver, spleen and bone marrow as well as in the lining of lymph channels in lymph nodes.

Right lymphatic duct: Collects lymph from the right side of the neck, right side of the chest, and right arm and empties into a blood vein in the right side of the neck.

Thoracic duct: The main duct of the lymphatic system that carries lymph upward through the thorax and discharges it into left subclavian vein.

Spleen: Purifies the blood; located near the liver, on the left side of the body in the area between the 9th and 11th ribs.

Thymus gland: Matures white T-cells; supports the immune system in fighting illness; located around the heart center. Many heart-centered exercises and meditations activate the thymus.

Study Points

1. What is a bacterium? What is a virus?
2. What activities might weaken the Immune System?
3. What activities might strengthen the Immune System?
4. What is a lymph node?
5. What is chyle?
6. Are there many types of white blood cells?

Chapter Twelve

The Endocrine System

Golden Glands, the Guardians of Your Health.

Basic Structure/Function

The Endocrine System secretes a variety of hormones. Hormones are secreted by glands and affect other cells in the body. Some hormones are specific and have local effects; some are general and affect virtually all glands. Two of the upper glands balance, orchestrate, and synchronize your glandular system: the pineal and the pituitary glands. At the pineal, pituitary, and hypothalamus area the mind, body, yoga, and meditation have maximum effect. This is the core control center and creates the brain soup, the blood quality, and the flow of consciousness.

Organs of the Endocrine System

The following glands release hormones and are activated by the hypothalamus or brainstem, which is like the control center of your body. The glands of the endocrine system are controlled by the autonomic nervous system and are released automatically when the body needs specific hormones.

Pineal Gland

The pineal gland is located in the center of the skull. The pineal gland has been considered by yogis to be the seat of the soul. In traditional medicine, its significance has been unknown, at best. What we do know is that, according to comparative anatomists, the pineal gland is a remnant of what was a third eye, located high in the back of the head of primitive animals.

The pineal gland plays a role in sexual cycles. It is regulated by light through the optic nerve and hypothalamus. The pineal gland secretes melatonin, which then stimulates the anterior pituitary gland that controls gonadotropic hormone. Animals that lose pineal function stop having annual sexual cycles. The pineal gland also plays a role in migration and other season-oriented physiologic adjustments.

The technology of covering the head with a cotton cloth provides a filter for the direct prolonged exposure to sunlight, which is a harsh experience for the pineal gland. The pineal gland can be overheated by direct solar exposure.

Pituitary Gland

The pituitary gland, also known as the hypophysis, is where the Nervous System and the glandular system meet and communicate.
It has 2 specific parts and 2 specific functions:

1. **The anterior pituitary** is known as the adenohypophysis and is glandular in nature. It is composed largely of glandular tissue, and manufactures and releases a number of hormones. Hormones of the anterior pituitary are:

    Growth hormone: Promotes growth by regulating protein metabolism.

    Adrenocorticotropin: Controls the secretion of adrenals and regulates the metabolism of sugars and fats.

    Thyroid stimulating hormone: Regulates thyroid hormone that effects all metabolisms.

    Prolactin: Milk production.

    Follicle stimulating hormone and luteinizing hormones: Regulates growth of the gonads and the reproductive cycles such as egg development and the menstrual cycle.

1. **The posterior pituitary** is known as the neurohypophysis and is neural in nature. It is composed mostly of neuroglia (nerve cells), and it releases neurohormones that it receives directly from the hypothalamus. Hormones of the posterior pituitary gland are:

    Antidiuretic hormone: Also known as vasopressin, controls the water concentration in the body through the regulation of kidney function.

    Oxytocin: Stimulates the delivery of milk during nipple sucking and aids in the delivery of the baby at the end of gestation.

The pituitary gland sits at the junction of the left and right optic nerve. The focus of the eyes specifically applies pressure to the pituitary gland. There are specific yoga kriyas and meditations that effect the pituitary gland.

Thyroid Gland

The thyroid gland is located in the front of the neck. Hormones of the thyroid gland are:
Thyroxine and triiodothyronine: Increase the rate of metabolism in all cells.
Calcitonin: Regulates calcium levels by promoting bone calcium absorption.
Iodine plays an essential role in thyroid function.
Hypothyroidism is associated with chronic fatigue, skin problems, depression, and weight loss.

Parathyroid

The parathyroid is closely related to the thyroid and is located just under the thyroid. It controls absorption, retention, and metabolism of calcium and phosphate.

Thymus Gland

Located around the heart center, many heart-centered exercises and meditations activate the thymus. It is involved in maturing white T-cells, and plays an important role in supporting the immune system and fighting illness.

Adrenal Cortex/Adrenal Gland

This gland is located on top of the kidney glands. Hormones of the adrenal gland are:
Cortisol: Regulates metabolism, associated with the stress response. See hypothalamic-pituitary-adrenal (HPA) axis.
Aldosterone: Regulates sodium concentration in the kidneys.

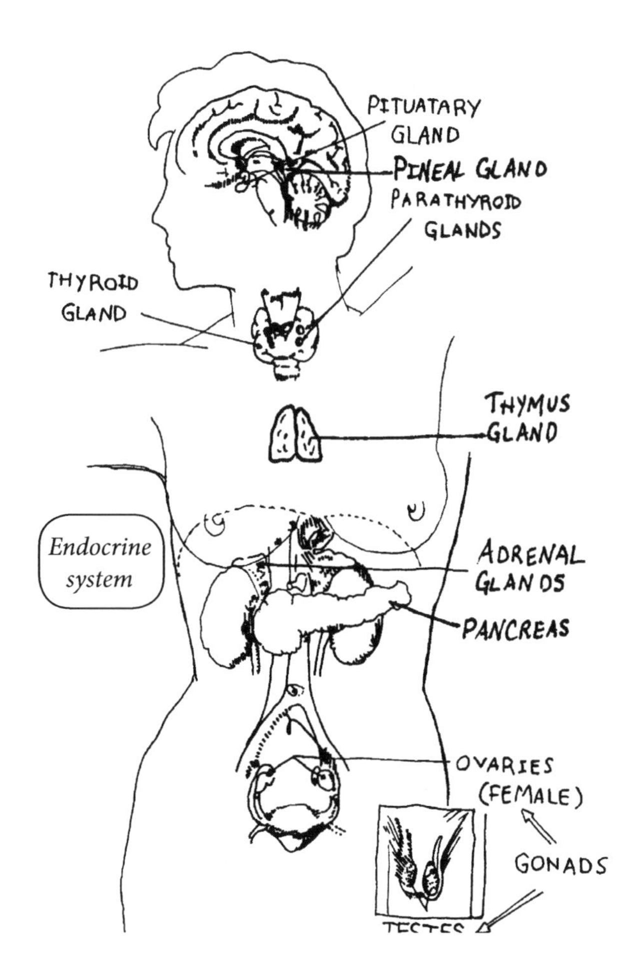

Pancreas

The pancreas produces insulin, which permits glucose to pass through cell walls to be used for energy and storage, and glucagon, which inhibits glucose absorption. Together they maintain the glucose level balance. The pancreas also contributes to carbohydrate, fat, and protein digestion.

The Gonads

The gonads regulate secondary sex characteristics and the development of the sex organs, ovulation, placental formation, milk production, and sperm production. In the female they are the ovaries and in the male they are the testes.

Ovaries

The ovaries produce estrogen and progesterone, controlling development of the ovarian follicle and thickening of the uterine wall.

Testes

The testes produce testosterone, which affects the sex drive in both males and females, and is involved in the development and maturation of sperm.

A Glimpse at the Relationship to Other Systems

The nervous system, the brain, and the glands are intricately linked within this neuroendocrine system.
The glands are the guardians of your health.
The glandular system has an effect on ALL other systems!

Negative Influences for This System

- Chronic stress
- Exposure to pesticides, which mimic hormones
- Prolonged diet of non-fat foods
- Hormones from animal products

Some Good Foods for the This System

- Onions
- Garlic
- Ginger
- Yogi Tea
- Proper mineral balance

Some Good Foods for the Pituitary and Pineal Glands

- Figs
- All "P" fruits
- Pistachios

Women and Their Moon Cycle

At the full moon: Try to drink only liquids on this day; milk only if necessary.
At the new moon: Try to fast on lemon-water only.
The eleventh day (after the full moon): Eat very sattvic this day; if you must eat, eat light and green things. If you just live on water and melon this day, your health will be optimal.
See The Master's Touch, p. 107.

How Yoga Helps

Yogi Bhajan: "What is Kundalini? The energy of the glandular system combines with the nervous system to become more sensitive so that the totality of the brain perceives signals and interprets them, so that the effect of sequence of the cause becomes very clear. In other words, one becomes totally, wholesomely aware. That is why we call it the Yoga of Awareness. Kundalini is the creative potential of a person."
Working on the chakras through stimulation of the Kundalini energy positively affects the health and balance of the glands. When the glands are balanced and healthy, it creates a sense of well being, by allowing the individual to feel the "connectedness" to all life.
The function of the pineal, pituitary, and hypothalamus is strongly impacted by the breath. Kundalini Yoga, particularly breath control, can greatly enhance the activity of the pineal gland. Focusing on the pituitary point (the third eye point at the brow) activates the pituitary (Kirtan Kriya is excellent for this).
Sound and mantra have a powerful positive effect on the glands, particularly the neuroendocrine centers of the hypothalamus, the pituitary, and pineal glands.

Yoga That Affects the Endocrine System

- Cold showers
- Massage
- Pranayama – slow deep breath, once per minute
- Breath of Fire

- **Adrenals**

- Beyond Fatigue Set (p. 18-20, Inner Workout)
- Exercises that twist and stretch the torso
- Pineal, Pituitary, Thyroid

- Wha Guru Kriya (p. 61-62, Intermediate Meditation)
- Yoga and meditation specifically for the pineal gland

Pituitary

- Ajnaa Stimulation Kriya (p. 23, Keeping Up With Kundalini Yoga)
- Pituitary Gland Series
- Breath Meditation Series for Glandular Balance (p. 72, Sadhana Guidelines)

Thyroid

- Any exercises that call for movement of the neck such as cobra pose, cat-cow, and shoulder stand
- Wahe Guru Kriya (p. 61-62, Intermediate Meditation)

Thymus

- Tapping on the chest
- Exercise sets for the heart center

Reproductive Organs

- Frog Pose
- Sat Kriya
- Camel Pose
- Tree Pose

Some Conditions of the Endocrine System

Diabetes: A disease which affects the insulin response.
Hyper or hypothyroidism: Too much or not enough thyroid hormone, which controls metabolism and kidney function.

Vocabulary

Endocrine system: The body system that includes internal organs that secrete hormones.
Hormone: Steroidal or amino acid-based molecules released to the blood acting as chemical messengers to regulate specific body functions.

Study Points

What is the function of the pancreas, pituitary, pineal gland, spleen and thyroid?

Chapter Thirteen

Yoga and The Body

13 Principles of Motion

Kundalini Yoga uses many different yoga techniques including physical postures. The 13 Principles of Motion will guide you to do yoga better and obtain the optimal yoga experience. In every Kundalini Yoga experience there is a dynamic tension, a still point, and a center of focus. This aspect can easily be missed. Focus on the intent as well as the mechanics. Guide the nervous system from your command center and the muscles, bones and ligaments will follow.

May these 13 Principles guide you to your path and may your journey along the way be enjoyable.

1. Breath

Allow the breath to do the work. If you concentrate on your breath during each exercise the movement will follow. The breath is crucial; it's your fuel. Increase the power of your breath, in any exercise and it will increase your focus, awareness, and stamina. Thus, in breathing intently, you will also decrease the chance of injury. Your body's ability is limitless, use the breath to allow the body to show you what it can do.

2. Gravity

Yoga developed for thousands of years here on Earth, it is therefore for people who live in gravity filled environments. Gravity is a big part of how yoga is done. There is a perpendicular force in physics called the Normal Force. You don't want to resist that force; you want to use it. The alignment of postures is so because of gravity. Have you ever seen a Ferris wheel? As one side of the Ferris wheel goes up the other side goes down. Frog Pose is a great example of this. When you're in Frog Pose you shift your weight forward on your fingertips and then the other end of the Ferris wheel (the buttocks) comes up. When you shift the weight back down on your heels, the buttocks come back down. You actually use the weight. The more you embrace gravity and the less you resist gravity, the more ease of movement you will have. In many yoga positions you almost defy gravity in this way.

The Law of Inertia is that objects that are at rest stay at rest; Newton's first Law is that objects in motion tend to stay in motion, and in a straight line, unless an external force changes the direction. You move the way you're used to, until you change the direction of the movement and then you can change the momentum for the movement too.

There is a dynamic tension in every posture in Kundalini Yoga. The Law of Reaction states that for every action there is an equal and opposite reaction and this law is how dynamic tension works. When anything moves, something else moves in a different way. In Kundalini Yoga postures, contradictory forces combine to create a single focal point; it is through this focal point that balance can be achieved in each posture.

3. Move from the Navel Point

Move from the navel point. Your core or natal nervous system is at your navel point. The actual navel center is 2 inches below the umbilicus (belly button). When you move and speak from here, you can be 100% behind the action of words and movements. "Every communication that does not come from the navel point brings disease, sickness, sorrow, sadness, madness, insanity, and bad luck" (Yogi Bhajan, p.18, Master's Touch,). This is true whether regarding communication with your movements or your speech. An implication of that is if you ever say things that are not in alignment with your essence, then you are reaching out and saying things that are not really you. If you can stay right behind your communication and are straightforward with yourself and others, there is no problem. Same thing is true with motion; whenever we move from any area other than our navel we create instability, which could create injuries.

4. Instantaneous Axis of Rotation

This principle of motion has to do with the spine. The idea is to keep the spine steady, straight and the vertebrae (spinal bones) stacked on top of each other, so that gravity transfers down to the ground for a base. In this way, the spine itself actually has less motion, which allows the rest of the body to move more, laterally (side to side) and in rotational ways (twisting). In every rotation there is one focal point that is not making an arc but is spinning on its own axis. The closer that axis is to the center of the spine, the more the spine will rotate on itself rather than wobble into an increasingly unstable movement. If you do not maintain a stable central axis, instability and an overall feeling of disconnection will prevail; each part of the body may feel only loosely connected to the next. When you maintain a stable and central axis you develop a secure core for rotational movement to come from.

5. Develop Deep Roots and Expansive Wings

A lot of Yoga is about how you plant the feet and/or hands to form an angle because this is how you attach yourself to gravity and interface with it. This can be called the roots. Deep roots establish a solid base and it is always wise to maintain it. The more grounded the roots the higher you can fly.

The wings are the lungs. Your wings are what carry you though the exercise. Usually there is a wing part of every position and a root part of every posture. It is a partnership that when functioning properly really help in getting the most out of any yoga practice. When you have deep roots the wings will move right and if you move the wings right, usually you can maintain the roots as well. Establish the roots and set the navel point before moving or positioning the wings. The stability of the navel point is directly proportional to the power of the breath. Begin correctly. End gracefully.

6. Be Aware of Clear Alignment

Wherever there is a restriction there will also be compensation--be it physical, mental, or spiritual. Restrictions of big joints (i.e. hips) result in pain in smaller joints. Areas of hypo-mobility tend to be compensated for by areas of hyper-mobility. For instance, if you have a hip problem it is just a matter of time before you have a knee problem. If you have a shoulder problem and you do a lot of triangle poses, the angle of the elbow will not be right, so you'll have problems with the elbow. Any rotation of the hip will create a knee problem or a problem with the ankle. Restrictions in the bigger joints will result in injuries in other smaller joints.

If there is an area of weakness, other areas will compensate by contracting and restricting movement in an attempt to stabilize the base. A great example is that the neck is plenty flexible but often the thoracic spine (the region of the ribcage) isn't flexible. Often times a person will compensate for this restriction in the ribcage by moving the head a lot. To do the posture in this way reflects a diaphragmatic tension that restricts the flow of energy in the body. Then the thoughts get stuck in the head. It's better to keep the neck steady than to create this illusion of success by simply moving the head. The idea is to lead with the heart and keep the chin in Neck Lock. Keep the heart forward and let the head follow. There is always this classical spiritual conversation between the heart and the head. The idea is, if the compassionate heart will lead you then the head will follow. If your head comes first then you don't listen to your heart so much.

7. Align All Aspects and Facets

Be aware of your thoughts and the architecture of the mind. Flexibility begins in the mind. If you start a posture correctly, with clear intentions, you will most likely end the posture correctly. A big part of keeping the mind focused is the specific dhristi (eye focus in a specific direction). Sixty percent of your energy goes to the visual cortex when you open your eyes; more so if you start to look around. Maintain awareness and focus by keeping the mind on "Sat" on the inhale and "Nam" on the exhale. Keeping the mind focused is crucial.

8. Strengthen what you want to Stretch and Stretch what you want to Strengthen

There is a relationship between being strong and being flexible. The physiology of muscles is very interesting. The more flexible you are, the wider range of motion a muscle has and the stronger it is. Overall the muscle is healthier because strength and flexibility also effect circulation. When a muscle is tight it has very little range of motion; not only is it's flexibility compressed but also the circulation and functionality is restricted. For instance, if you are going to pick up a pencil you have to bend down 180 degrees. If some of the muscles that are suppose to move ten degrees only move five degrees that difference has to be made up somewhere else. Areas that are less restricted will compensate for areas that are more restricted. Strength and flexibility are crucial for yoga.

9. Smile

The goal of yoga is happiness! Reflecting the goal on your face firmly establishes your desired outcome. When you are doing yoga, a smile will always help you to retain awareness of your comfort level. Maintaining this facial mudra will let you know when you are using too much effort or when you have created excess discomfort. If you are grimacing and forcing yourself through the posture or kriya, you're trying too hard. If this happens, breathe more, smile, and go through the process. Even if you're not in a good mood, try smiling a few minutes. Smiling while immersed in your yoga practice will give you a different perspective on your happiness or unhappiness. There may be a mudra and asana your practicing, and if you can remember your smiling facial mudra, your mood will always be one that is elevated.

Bonus: Smiling also stimulates the immune system. When the cheeks go up the immune system is invigorated, it's a big deal.

10. Keep the Outcome in Mind

The purpose in practicing yoga is to attain the attributes of a yogi: humility, focus, radiance, integrity, and neutral-mindedness. You know yourself very well. You know your blocks and the aspects of yourself you are working on. If you keep those things in mind when you're doing yoga you'll find success, you'll project, and you work through your patterns towards your goal. If your goals are nebulous, you just want to get "somewhere," it's a lot harder. Luckily, however, and even if your goals are not clear, you will still certainly benefit from your practice, something will happen.

Cells talk to each other all the time and one of the reasons that we can transform so well as humans is that cells turn over frequently. Some of the oldest cells in your body are just 3 months old. Every three months, all the bone cells turn over. Osteoclasts eat up old bone cells and osteoblasts put down new bone. Our skin cells are a couple weeks old; the stomach cells inside the lining of the stomach are six minutes old. Don't think you can't change? You change anyway, all the time. As you reinvent yourself, your cells continue to divide. Utilize your command center. The only time is NOW. The past is over, and the future is pure fantasy. Let go of what is not the essential You. Move beyond the polarities and into the blissful abyss.

11. Vitality Naturally Becomes Virtue

Vitality is a natural state of energy, enthusiasm, and focus. A strong sense of vitality leads to the virtuous state of doing the right thing and attracting positive environments as well as embracing trust. A non-vital state is a state of weakness, a state of neediness and often the state of poverty.

Radiant health, abundant wealth, and a positive projection are all reflections of vitality. Virtuousness is moral and ethical excellence. A strong Navel Point (vitality) leads to an open heart (virtue).

12. Commitment Creates Momentum

To be successful in any yoga practice, you must have 100% committed breath and that means 100% on the inhale and 100% on the exhale. As soon as you start to breathe unconsciously, your mind goes unconscious, then your body is kind-of in the exercise but by then your intentions are definitely not. At this point you lose the exercise. You lose your awareness. Focus. Seize the Day! Most exercises are 3 minutes or less. Get the maximum effect per second. Let go of your patterns, stay focused, and breathe through the exercise. Do your best, with every breath, in every exercise, and in every kriya. Juice the moments. Allow the Now to happen. Let go of the past, and forget the future. Be here now.

If you are having trouble focusing in a kriya or a posture for several minutes try doing it for one minute at a time. One minute where every cell is committed to that one minute of doing that exercise for that moment. One minute, one complete minute, one totally focused minute, and then lay down for 15-30 seconds. Repeat as necessary.

13. Transformation is the Path

We tend to fold on our preexisting creases. Initially, doing yoga correctly does not make it less difficult, although it does make it more effective and more productive; and in the long run, doing yoga correctly does make it easier. It is easier to use gravity, it is easier to use the breath, and you may find it easier to change your patterns. So your head might not touch your knee any more, but this is no longer your goal. There are a lot of ways your body may fold on itself that you may be attached to, because humans identify with their patterns (limitations). As you work through yoga postures, you will move through your limitations and it will give you a different perspective on you. When you change certain habits in the way you live (like your posture, your breath, and how you walk and talk etc.), then you begin to change the core of you. You begin to see things very differently. You may see that the goal of yoga is not to get your head to your knee. Intention goes a very long way.

YOU ARE THE GOAL

Yoga is an inner journey. YOU are the goal. Be who you want to be, not who you were yesterday! Act with the outcome in mind. Every motion takes you closer to the goal of knowing the Infinite Self. You are unlimited by nature. You are the universe. There is nothing to "get" that you don't already have. Control the breath. Quiet the mind with the mantra and Shabd, and experience the Soul, the light of 10,000 Suns.

CHAPTER FOURTEEN

THE LOCKS

The locks (bandhas) are used to focus and direct prana. They direct the heat of the psyche (Tapa) into the main nadis. The three major nadis are the Sushmana (central), Ida (left), and the Pingala (right). They are located in the spine. Tapa is needed to awaken the dormant Kundalini energy. This is your true essential nature.

ROOT LOCK (MULBANDH)

The first lock is a collaboration of the lower three chakras. It is the contraction of the muscles of the anus/rectum, sex organs, and the Navel Point. The center of the Navel Point is located about one to two inches below the bellybutton.
Root Lock consolidates the energies that travel through the nadis. First the anal muscles are contracted and the urethra is retracted. This mighty force is consolidated and condensed as the Navel Point is pulled back towards the spine. As Mulbandh is applied, positively charged prana mixes with the negatively charged apana at the Navel Point. This increases heat in the area of the 4th vertebrae, activating the Kundalini energy. Prana is energy that is taken in, like breath and food. Apana is the energy of elimination. Between intake and output is digestion. The more efficient digestion is within the body, the easier the elimination. Just as everything that goes up must come down, everything taken in must be "Let Go." Unimpeded nadis are essential for good physical health, mental clarity and spiritual stability. Utilization of Root Lock and the Navel Point during all physical activities is a key to the health of a yogi.

DIAPHRAGM LOCK (UDDIYANA BANDH)

The diaphragm is a "U" shaped muscle located at the base of the ribcage. When we inhale, the diaphragm expands down, and as we exhale it rises up underneath the lower part of the ribcage. Diaphragm Lock involves a maximum lifting of the diaphragm muscle, creating a negative pressure in the lungs. As all of the air is powerfully pushed out, the pressure in the lungs becomes significantly less than the pressure of the outside air. It is this partial vacuum that creates an opportunity for the next complete inhale. Diaphragm Lock and diaphragmatic movement is central to the flow of prana which fuels all yoga.

NECK LOCK (JALANDHAR BANDH)

The major conflicts of life are between the head and the heart and often manifest in the neck. Neck Lock straightens out the unique and natural curves found in every neck. The lordotic (forward) curve straightens, and the occiput, the back of the head bone, is lifted. The positions

of the neck and jaw are closely related. Retraction of the jaw is a significant part of Neck Lock; the jaw is pulled posterior (back) to lift the structures of the throat and the back of the tongue. At the top of the sternum (breast bone) there is a notch. During Neck Lock, this notch will retract due to the negative respiratory pressure as well as the elevation of the glottis (flap at the back of throat). By these slight retractions, optimal alignment is achieved.
There is a sensation in the head unique to the Neck Lock experience. As Neck Lock is applied, the neurological flow and dynamics of the fluid are affected. The major nerves and the cerebrospinal fluid flow in the central and posterior parts of the spine. As blocks in the neck are relieved, blood and cerebrospinal fluid flow freely.
When practicing Kundalini Yoga a lot of energy is generated and circulated throughout the body. The precise slight adjustments made in applying Neck Lock allow the Kundalini energies to flow effectively.

It is important to apply Neck Lock during all meditations.

How to do it:

Sit in a comfortable meditative posture such as Easy Pose.
Straighten and elongate the spine. Slightly retract the chin in, and push the top of the head up towards the sky. There is a perfect place to tuck the chin too; you feel uplifted and aligned when you find the spot.

The Great Lock: (Mahabandh)

Mahabandh is the simultaneous usage of the Root Lock, Diaphragm Lock, and Neck Lock. It has a systematic rhythm. First is the pulling up of the Root Lock. This activation creates momentum resulting in the lifting up of the diaphragm, Diaphragm Lock, and the retraction of the neck, Neck Lock. The elongation of the entire spine sends the eyes to focus up at the top of the head. Mahabandh is composed of the three locks, blended into one action, one motion, and one intention. Mahabandh energizes the sexual and sensual systems, feeding the nerves and glands with supercharged prana. It is recommended to relax or meditate after the application of Mahabandh, unless otherwise specified.

Chapter Fifteen

The Power of Prana

Ida, Pingala, Sushmuna and more...

As humans we have 72,000 nadis. Most of them originate from the natal nervous system and radiate out from the navel center. These are the pathways of energy that flow through the body. The nadis are hardwired into our matrix.
Of these 72,000 nadis there are three central channels, the ida, the pingala and the sushmuna. These channels run along the spine and are woven within the chakra system.

Ida

Ida is the receptive pathway and controls the energy on the left side of the body. This is the listening, feminine, calm, receptive side of the human.

Pingala

Pingala is the projective pathways of the body on the right side. This is the masculine active side of the human.

Sushmuna

Sushmuna is the central, center and core pathway. It neither receives, nor projects. It vibrates at the primary frequency of life. It is the brainstem and the deep structures of the cord. It is the central nervous system. It is expressed as the primal wave of fluid in the spine and is directed by the frontal lobe of the brain.

Prana

Prana, the fuel of life, flows through these channels. It is an essential life force that feeds the nerves, organs, and tissues of the body. It gives the mind the energy to focus and meditate. Prana is similar to chi and the nadis are similar to the meridians. The flow of prana, through the nadis, creates a highway of electrical energy that feeds the body and keeps the mind vital. The motion of prana throughout the body and the mind helps us to feel charged and full of energy. When we are full of prana, we are ready for life. The body's respiration is open and the mind responds positively to the surrounding environment and we are able to sense the motion that is in the entire universe. When we are personally empowered, we radiate compassion.

Vayus

There are five subdivisions of this pranic life force, and these are called vayus. Inside the body, they occupy specific regions. We will look at these five regions one by one, starting from the bottom and working our way up.

Apana

Apana vayu is located in the intestinal region just below the navel point governs all functions of elimination. Its energy moves down and out. With strong apana, we feel a sense of security and the ability to walk with confidence. An imbalance will cause us to feel slow and confused.

Samana

Samana vayu, is located between the heart and the navel, is responsible for the body's metabolic activity. It is in charge of organizing and systematizing. When samana is balanced, we are able to discriminate and have strong emotional clarity. An imbalance will result in mental and emotional confusion resulting in the inability to discriminate what emotions are being felt. There is also trouble maintaining healthy boundaries.

Prana

Prana vayu is located in the chest region between the heart and the neck, is connected with the lungs and governs inspiration. The motion of prana vayu is accumulation and expansion of the lungs and of the internal energy. The mind takes on a positive outlook on life when this region is balanced.

Udana

Udana vayu is located at the larynx and upward in the head, interacts with our projective nature. Its motion is upward and out. Everything from speaking to vomiting is controlled by udana, and when there is balance here, one is able to project and create with the spoken word. An imbalance causes shyness and the inability to correctly make musical pitches.

Vyana

Vyana vayu governs the whole body's coordination, movement, and integrity of thoughts and sensory awareness.

An imbalance in one of the vayus directly corresponds to an imbalance in one of the Ten Bodies.

The Five Tattvas/Elements

We are made up of five tattvas, the elements. Our physical well-being is maintained by these five earthly tattvas.

Earth

Earth (Pruhivi) is matter in its solid phase. Shape, form, stability are all components of earth. The structures of life such as bones, cartilage, hair, nails, muscle, tendons are all created from the earth element. Without the earth element, our bodies would lack the necessary structure and resistance to the forces of nature. In essence, the earth element cradles and holds all living creatures of the planet, providing food, shelter, and roots.

Water

Water (Apas) is matter in its liquid state, acting as both a carrier and agent of change. It represents our blood, carrier of oxygen, our saliva and digestive fluids, our hormones, our sweat and our urine. Water is the mover of nutrients and the eliminator of waste products. It is the sea of life within us. The water element is vital for assimilation and for maintaining electrolyte balance. While the body is composed of more than 70% water, the plasma in our blood is made up of more than 90% water. The body's lymphatic system is also governed by the water element. Water helps all the other elements maintain their function. Even the fiery digestive acids inside the stomach requires water to function. The element of water creates the connective flow between breath and movement.

Fire

Fire (Agni) is the energy of transformation and change. The exchange of energy, the conversion of matter from one phase to another, are both manifestations of the fire element. In the body, fire governs metabolism (conversion of food into energy), body temperature, and even thoughts (assimilation and digestion of sensory impression into cognitive understanding, reason, and memory). Where there is movement there is friction, and where there is friction, there is fire. This element is radiant energy and is present in the body as the flame of attention.

Air

Air (wind) (Vayu) is matter in its gaseous phase. It is movement, fluidity, and formlessness. It can be compressed or expanded. It can be a gentle breeze or a forceful gale. Air manifests itself in the body as movement, pulsation, expansion, and contraction. Air cannot be smelled, tasted or even seen. It can only be felt through the sense of touch. Prana is the basic principle of the air element, and flows as the consciousness from cell to cell in the form of intelligence. It is our vital life force. Movements of the heart, breathing, intestinal peristalsis, and other involuntary movements are governed by the air element.

Ether

Ether (Akasha) is a mystical concept meaning all enclosing, all pervasive, omniscient, omnipresent. Akasha is the emptiness in which all other elements exist. Quantum mechanics defines ether as "the field." It represents the nature of the void, or the "home" in which all other elements interact. To grasp this concept, consider a doorway or the empty space inside a vase. There is nothing there, yet in that empty space lies its usefulness. Similarly, the space in your body, the lungs, the nostrils, the mouth, the stomach, are all useful because of the space within them. Space implies receptivity, distance, and location. In our bodies ether is the first expression of Consciousness, and the basic need of all bodily cells. In the development of matter, ether comes first. Ether is expansive, empty. It has no resistance and provides freedom in which to move. Without the ether element, there is no love or freedom.

The Doshas

In ancient times, food was considered medicine. People used food and herbs as a way to purify and correct imbalances in the body. Today, we are slowly coming back to this concept, recognizing that "we are what we eat." According to Ayurveda, there is a state of inner balance between the three doshas called vata, pitta, and kapha. These three forces exhibit qualities unique in every person, and assist in maintaining the physical and mental bodies. These energetic factors also correspond to the tattvas (elements), and can be recognized by the following attributes:

Vata Dosha

Vata Dosha helps to maintain the body and is the source of all physical movement. It controls the mind, senses, and causes the elimination of wastes. The elements linked with this dosha are air and ether.

Pitta Dosha

Pitta Dosha helps to regulate the inner heat and digestive fire. It also perpetuates the formation of blood. The elements linked with this dosha are fire and water.

Kapha Dosha

Kapha Dosha is responsible for nourishing and lubricating the body. It helps to maintain sexual potency and sustains the mental balance of the individual. The elements linked with this dosha are water and earth.

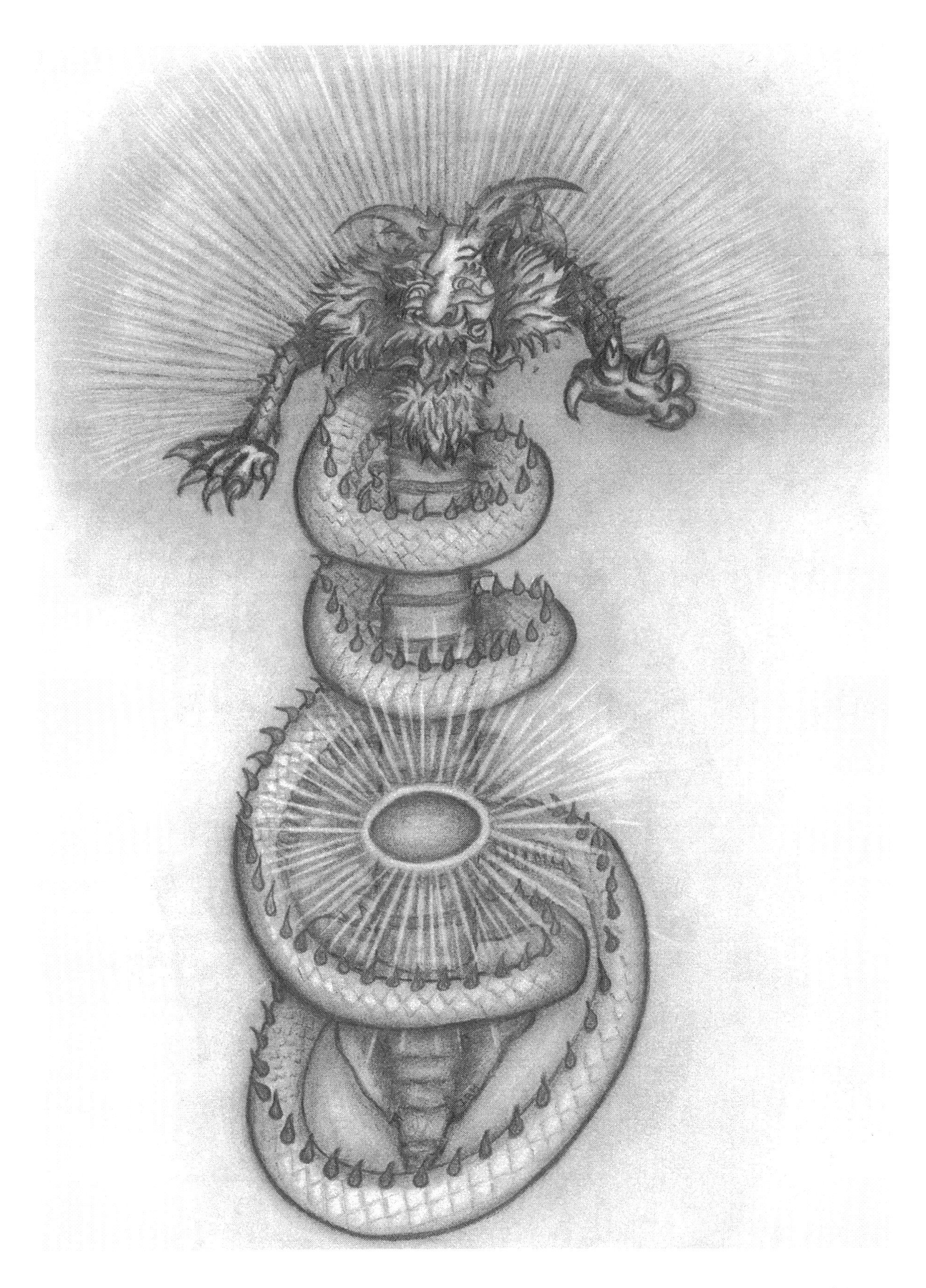

Chapter Sixteen

The Chakras

Our body is made up of several trillion cells. Each cell develops in a specialized way to become a specific type of tissue. Tissue becomes organs and is coordinated through the nervous system. Groups of organs work together like a team to perform the functions of elimination, digestion, reproduction, respiration, circulation, etc. One way to understand these systems is through the basic anatomy of the physical body. Another way to understand this concept is through vital energy centers known in yogic terminology as chakras. Derived from the Sanskrit language, chakra means "wheel." The energetic quality of a wheel is circular. Each chakra is experienced through a nerve plexus that governs that area. When our chakras are functioning optimally, we experience ease in our ability to concentrate, communicate, and project. An imbalance in any one of these energy centers can lead to a long list of ailments and disturbances in the body, both temporary or long term. Through consistent effort to clear, rebalance, and energize these energy centers we can prevent chronic illnesses, habitual patterns of behavior, emotions, and thoughts.

The chakras are interconnected with an energy pathway known as the sushmuna, or "spine." The prana, or energetic life force, travels in a serpentine manner up the sushmuna from the First Chakra up to the Eighth Chakra. The internal structure of our chakras originates from the nature of the resonance at that chakra.

The chakras are specific areas in the body that hold specific frequencies of life that are manifested through a particular cluster of nerves that control organs and tissues. These specialized areas function to vibrate and project this energy. These illuminated pathways are our access to the subtleties that hold the joys of a life of awareness.

Below are further descriptions of each of the Chakras, the designated Tattvas or elements, related to each chakra, their location in the body and how they relate to the physiological functions of all the biological systems within the body. Let this serve as a guide for you to have a solid understanding of this ancient system and its inherent wisdom.

The Lower Triangle

First Chakra: Mulhadhara

Action: Gravity	Mantra: Let Go	Tattva: Earth

The First Chakra is located at the base of the spine, between the anus and the genitals. It is connected with the sacral-coccygeal nerve plexus and associated with the excretory system. When this chakra is strong and balanced, we feel centered, grounded, stable, and secure. An imbalance causes fear, feelings of isolation, attachment, an overall weak constitution, and mental fatigue. The element connected to the First Chakra is earth.

Second Chakra: Svadisthana

Action: To Create	Mantra: Create	Tattva: Water

The Second Chakra is located at the sexual organs. It connects to the nervous system via the sacral nerve plexus. The reproductive glands, sex organs, bladder, and kidneys are associated with this chakra. When this chakra is balanced, creativity, calmness, patience, and positive relationships occur. If this chakra is under stress, we experience magnified fluctuations of emotion, increased feelings of guilt, and reproductive and kidney malfunction. The element connected to the Second Chakra is water.

Third Chakra: Manipura

Action: Transform	Mantra: I am, I am	Tattva: Fire

The Third Chakra is located at the Navel Point. The solar plexus governs the Third Chakra. The organs that are linked to this chakra are the adrenals, gall bladder, liver, spleen, pancreas, navel plexus, and digestive organs. A balanced Third Chakra results in a sense of self-empowerment, commitment, healthy self-esteem, and inspiration. A stable Navel Point connects us with our infinite identity. When this chakra is out of balance, anger, shame, and greed often occur. A lack of spontaneity, general weakness, bad timing, and poor digestion are also typical. The element connected to the Third Chakra is fire.

The Bridge

Fourth Chakra: Anahata

Action: Love	Mantra: Compassion	Tattva: Air

This chakra is considered the balance point between the lower triangle and the upper triangle. It is located at the heart center (between the nipples). The cardiac nerve plexus governs this region. The Fourth Chakra energizes the blood and physical body with life force. The organs governed by this chakra are the heart, lungs, and thymus gland. Healthy functioning of this energy center manifests as compassion, harmony, a neutral mind, spiritual awareness, and the recognition of others. An imbalance results in jealousy, possessiveness, controlling behavior, grief, dependency on others for love, and a fear of rejection. The element connected to the Fourth Chakra is air.

The Upper Triangle

Fifth Chakra: Vishuddha

Action: Communicate	Mantra: Vibrate the Cosmos	Tattva: Ether

The Fifth Chakra is located at the throat. The pharyngeal plexus governs this region. Organs and structures linked to this chakra are the cervical vertebrae, trachea, vocal cords, throat, parathyroid, and the thyroid. When this energy is balanced, we have the courage to project with spoken word, communicate freely, and positively express our authentic self. When the Fifth Chakra is imbalanced, we are sluggish, shy, challenged in verbal expression, and are afraid of being harshly judged. Neck, throat, and thyroid problems are present. The element connected to the Fifth Chakra is ether.

Beyond the Tattvas

The Sixth through the Eighth Chakras are beyond the Tattvas/elements. They correspond to more subtle realms. One way to understand this is through the quality of light, with varying spectrum, which is quantified by frequency of wavelength and intensity.

Sixth Chakra: Ajna

Action: Receive and Perceive	Mantra: Understand	

The Sixth Chakra is located at the third eye point. (1/2 inch above the eyebrows at the middle of the forehead). The hypothalmaus-pituitary axis governs this chakra. The organs associated with this chakra are the pituitary gland and the brain. The pituitary gland is where the nervous system and the glandular system meet. When this chakra is balanced we are able to experience devotion, clearly perceive our intentions and purpose, have increased intuition, and the determination to project our innate wisdom. The Sixth Chakra specializes in recognizing, analyzing and understanding patterns. When there is a blockage, we feel confused, depressed, and over-analytical. We are unable to focus on a given task.

Seventh Chakra: Sahasrara

Action: Connect with Infinite	Mantra: Be One	

The Seventh Chakra is located at the crown of the head. The central nervous system governs this region. It is associated with the brain and the pineal gland. When it is balanced we are humble, connected to our higher self, and open to the unknown. We experience a feeling of union and bliss. When this center is unbalanced, we have a tendency to feel doubtful, fear death, and experience a general feeling of separateness. This chakra connects us with cosmic sound. When it is open and clear, we become one with universal sound. Here is where truth and reality crystallize into a clear experience of the matrix of Infinity.

Eighth Chakra: The Aura

Action: Radiate	Mantra: Luminate, Vibrate and Radiate	

The Aura, also called the electromagnetic field, is the sum total of the electric field of all of the cells. It is the radiance at our core that beams out into the infinite. A strong aura manifests in a sense of oneness with the universe and compassion for the self and all other beings. It is apparent in a strong and positive presence. It also helps to filter out negative energy. It extends 9 feet in all directions. When the electromagnetic field is weak, there can be a tendency to feel withdrawn, vulnerable, and closed to one's surroundings. The nervous system is sensitive to changes in the electromagnetic field. This chakra is associated with a sense of being.
One way to think of the chakra system is through the quality of light and the many frequencies possible in its expression. Slower waves of light are radio waves and microwaves. Faster waves are infrared and visible light. The fastest waves are x-rays and gamma waves. Ultra-fast waves are the cosmic waves that prevail in every area of the universe. They are so fast and small, they go through us constantly. We relate to our world through our sensory experiences, so visible light is often used as a more tangible model. The frequencies of light start at the slower

such as red, and progress to orange, yellow, green, blue, and purple. All the models are ways to describe an internal experience. They are all pure fantasy, a way to share and save an experience.

Chakras

	TATTVAS	PHYSIOLOGICAL SIGNS OF WEAKNESS	POSTURES TO STRENGTHEN EACH CHAKRA
1. Muladhara (Root)	Prithivi (Earth)	General weak constitution, problems with elimination, decreased physical and mental resistance	Chair Pose, Crow Pose, Frog Pose, Mulbandh,
2. Svadhistana (Sacral)	Apas (Water)	Kidney and reproductive imbalance, irresponsible relations, rigid emotions	Cobra pose, Butterfly, Sat Kriya, Cat-cow, Pelvic Lifts
3. Manipura (Navel Solar Plexus)	Agni (Fire)	Digestion, gall bladder, liver, and pancreas imbalances	Stretch Pose, Peacock Pose, Bow Pose, Breath of Fire
4. Anahata (Heart)	Vayu (Air)	Heart and lung imbalance, blood pressure abnormalities	Ego Eradicator, Bear Grip, Baby Pose, Yoga Mudra. All Arm Exercises, Upper Torso Twists, and Pranayam
5. Visshuda (Throat)	Akasha (Ether)	Throat and neck problems, thyroid imbalance	All Chanting, Shoulder-Stand, Plow Pose, Camel Pose Cat-Cow, Neck Rolls, Neck Lock, Nose to Knees.
6. Ajna (Third Eye)	Beyond the Tattvas	Depression, over-intellectualizing, confusion	Meditation (on 3rd Eye), Archer Pose, Kirtan Kriya, Whistle Breaths, Yoga Mudra, Long Chant. All Postures with the Forehead on the Floor.
7. Sahasrara (Crown)	Beyond the Tattvas	Separate from existence, lack of abundance, fear of death	All Meditation, Ego Eradicator, Mahabandha, Sat Kriya, Eye Focus at the Tip of the Nose.
8. Aura (Electromagnetic Field)	Beyond the Tattvas	Grief, Fear of Death, feel separate from existence and abundance, feeling closed to surroundings, withdrawn, vulnerable, shy	All Arm Movements and Meditation. Triangle Pose, Archer Pose, Ego Eradicator

Chapter Seventeen

The Ten Bodies

The Ten Bodies and the Eleventh Embodiment

It is a common misconception to believe that we have only one body. In actuality, we have Ten Bodies. All of the Bodies are equally present, real, and have a profound impact on our wellbeing. We have one physical body, three mental bodies, and six energetic bodies. The Ten Bodies are projections of the psyche. The psyche is an aspect of the mind. The psyche is defined as many things, from a character in classical Greek mythology, a female name, an aspect of psychology or the soul, spirit, or mind. To avoid falling into the new age psychobabble arena, allow us to define psyche as an aspect of the brain that we call the mind. The psyche contains the soul, the mind and active form of soul: the spirit. The brain is the controlling aspect of the nervous system that commands the entire being and contains the projection of the self we call the psyche. The psyche is the projection of the mind and spirit. The Ten Bodies are aspects of this inner organization, leadership, and projection.

As we move through our lifetimes, the psyche maintains a focus on the individual's destiny, or the ultimate path of their soul. Certain bodies play more dominant roles in different lifetimes; depending on what lessons the psyche is working on in the effort to live its' destiny.

The Ten Bodies are each unique. They carry great potential for manifesting specific energies. They also have certain tendencies that occur when they are weak. Fortunately all of the Bodies can be strengthened with the practice of Kundalini yoga.

The Ten Bodies

The Ten Bodies are chakras of the mind, aspects of mental projection compartmentalized carefully to correspond to mental tools we use daily, and become aware of during meditation. They are specific frequencies of consciousness. These ten aspects of the psyche correspond to aspects of the mind and functional aspects of the body. The chakras are a map of physical functions beginning at the base of the spine and going up to the top of the skull. The Ten Bodies connect the physical aspect with the mental aspects of the self. Together the Ten Bodies and the chakras make up a matrix of the hue-man being. The practice of Kundalini Yoga and Meditation accesses the hidden gifts inside untapped talents, and undiscovered truths.

The First Body: The Soul

Heart Over Head	First Chakra	Tattva: Earth

This is your most essential self. The core. The part of you that is neither your body nor intellect. The soul is vast, unlimited and illuminating. It is timeless. The soul is very individual and a piece of something much larger; the totality of all souls and all that is or is not.
When you approach life from the perspective of the soul you develop a context of compassion. Within this context, the mind naturally becomes contemplative. The control center of the self is this soul and if the signal is clear to the body and the mind, there is a manifesting innate type of intelligence that permeates, organizes, and regulates the functions of the body and mind.
We are spirits (souls) here to have a human experience. We are not humans seeking a spiritual experience. There are lessons in being a human being that are essential to spiritual development. Kindness, compassion, and truth are the frequencies of the soul. ALL is ONE, without duality is the soul's mantra. The soul body has a direct relationship with the First Chakra.

The Second, Third and Fourth Bodies

The next three bodies are the major subdivisions of the mind: Negative, Positive and Neutral. Together they comprise the mind which is an aspect of the central nervous system. The Negative Mind has qualities similar to that of the ida. The Positive Mind has qualities similar to that of pingala. The Neutral Mind has qualities similar to that of the sushmuna.
At the top of the head is a thousand petal lotus flower. Each petal is a thought, and these thoughts occur in a split second (in the blink of an eye). The thoughts flow through neuro-pathways into different regions of the brain, depending on the nature of the environment as well as your innate and educated mind.

The Second Body: The Negative Mind

Longing to Belong	Second Chakra	Tattva: Water

The Negative Mind is the protective mind. The mantra for the Negative Mind is, "from emotion to devotion." The projection of the Negative Mind is "longing to belong."
The Negative Mind has the ability to see the possible negative outcome of any situation.
The Negative Mind is the mind of the lawyer that you pay to read all the contracts, overlook the paperwork, and keep you out of trouble. The Negative Mind has a desire to develop successful, meaningful relationships, and understands the risks involved. When the Negative Mind is functioning optimally you can balance the risk and rewards of any situation. The second mind listens to the emotions and is governed by the water tattva. The Negative Mind must embrace discipline (Saturn energy) to obtain success. The second body has a direct relationship with the Second Chakra.

The Third Body: The Positive Mind

Devil or Devine	Third Chakra	Tattva: Fire

The Positive Mind is aware of opportunity. It loves expansion. The Positive Mind has momentum, creates vision, and develops leadership. The Positive Mind is governed by the fire tattva and has a direct relationship with the Third Chakra. The Positive Mind is a natural flow of optimism. The mantra of the third body is Devil or Devine.

The Fourth Body: The Neutral Mind

Cup of Prayer	Fourth Chakra	Tattva: Air

The Neutral Mind opens the heart. The Neutral Mind balances internal forces including caution and opportunity, Jupiter and Saturn, expansion and retraction, saving and spending, and positive and negative. When the Neutral Mind is activated you approach every situation with maximum sensitivity and with NO assumptions and preconceptions. A fresh perspective gives your mind the chance to see without filters, without a personal perspective, without a position. Free from personal position you have a chance to see the truth, if you can become clear of yourself.

The Fifth Body: The Physical Body

Teacher/Balance	Fifth Chakra	Tattva: Ether

The Fifth Body is the command center of the neuro-endocrine system. It is the body of the teacher. The Physical Body is a complex group of cells that works together as a vessel for the soul. The physical body has many lessons to teach us. The Physical Body takes a long time to manifest in terms of spirit. It takes 8.4 million years to manifest the spirit's spiritual development enough to exist for 1 light year. It takes 100,000 light years to obtain (rent) a body that is capable of enlightenment (a self awareness).

In total, you are not your Physical Body, however without it, you cannot have human experience, so it is rather important to the years you live and the life in those years. The Fifth Chakra is directly related to the Fifth Body. The cervical plexus neurologically innervates the Fifth Chakra. Health is regulated from this region, and in ancient times was known as "the mouth of God." The healthcare professionals who focus on this area are Upper Cervical Chiropractors. They focus on the pathway of the nervous system as it passes through the Fifth Chakra.

Physical health is an expression of intelligence that is innate to existence. The nurturing and support of these natural self-healing processes are an essential aspect of wellness, spiritual evolution, and the optimal practice of Kundalini Yoga.

The Sixth Body through the Eleventh Embodiment are Beyond the Tattvas

The Sixth Body: The Arc Line

Power of Prayer	Sixth Chakra	

The Arc Line is an aspect of the electromagnetic field. It is the interface between inner vision and outer manifestation. The Arc Line is about integrity and consistency. Your most inner manifested frequency becomes your outer reality. In other words, what you think about most, becomes what your life is about. Your most common thought becomes your most manifested reality. When your Arc Line works well you are clear, crisp, spontaneous and inspired. When it is not working you will feel frustrated that your vision and leadership is not successful. This is the region of prayer and attraction.

The Seventh Body: The Aura

Platform of Elevation	Seventh Chakra	

The Aura/electromagnetic field is a measurable electric force that surrounds living organisms as well as planets and electrical devices. Much of our biology is controlled by this electromagnetic field. The Yogi is acutely sensitive to the electromagnetic field. When your Aura is strong you feel confident and secure. Your very presence can uplift others. If your Aura is weak you can feel like a satellite dish, sensing the pain of others so much so that you do not feel separation between yourself and them.

The Eighth body: The Pranic Body

Finite to Infinite	Eight Chakra	

This is the energy management system of the self. It allocates and regulates energy. Sometimes the Pranic Body gives energy direct from breath, like during meditation. At other times energy is stored in the Pranic Body in places like the prana cavity, under the diaphragm. And other times energy catalysts from outside is used to modulate the inner flow of energy.

The Ninth Body: The Subtle Body

Mystery or Mastery

This is the carriage of the soul, the door to the subtle realms and the crown of meditation. The Subtle Body reads between the lines and knows the unknown. The Subtle Body is developed after 1000 days of practice of a specific discipline, a minimum of one hour a day. This opens the door to the experience of mastery. Mystery or mastery is the motto of the Ninth Body.

The Tenth Body: The Radiant Body

All or Nothing

The Radiant Body provides a platform of commitment and consistency for positive self-image and self-value. Everyone has an aspect of himself or herself that is a leader, crowned with glory, a royal and regal crowned prince or princess. This is the Radiant Body. The glow, vitality and purity that you get when you practice Kundalini Yoga is the Radiant Body. Dressing in elegant style, moving with grace and accepting the gifts of life are all expressions of the Tenth Body.

The Eleventh Embodiment: Infinity

Beyond

The Eleventh Embodiment is a totality of all the bodies and a little more. It is that extra something special, that Bonus and that which is more than the sum of the parts. It is the essence of pure sound current from which all mantras originate. It is a connection to Infinity. There are several indications that the Eleventh Embodiment is working well. We stop struggling and trying and allow the resources of Infinity to flow through us. The resources of Infinity are perfect timing, innate intelligence, and a state of non-duality. This occurs when there is a pure experience of reality in the present moment. This requires a letting go of pre-conceptions, intellectual rationalizations, and the psychological web of drama or finite personal identification. The Eleventh Embodiment is a state of non-dual openness and ecstasy. We can deepen the awareness of our own identity if we clearly understand the attributes specific to each of the Ten Bodies and the Eleventh Embodiment.

Ten Bodies				
BODY	**DESCRIPTION**	**KEY PHRASE**	**KEY QUESTION**	**PERSONALITY TYPE**
1st body: Soul Body (Inner Self)	This is the body of spiritual leadership. It is the true destiny of your divine life.	Heart over Head	Can I be compassionate and creative with myself and others in my actions and attitude?	Holy Person, Wise One
2nd body:Negative Mind (Protective Mind)	This is the body of devotion. It turns your emotion into a stronger devotion	Longing to Belong	Can I calculate the danger in a given situation and have clear boundaries?	Lover
3rd Body: Positive Mind (Projective Mind)	This is the body of potential. Opportunities are limitless when a clear vision comes together with this energy.	Devil or Divine	Can I allow myself to be hopeful and experience the wonders of life?	Creative Nurturer, Entertainer
4th Body: Neutral Mind (Meditative Mind)	This is the body of neutrality. It is possible to see both the positive and negative in all situations.	Cup of Prayer	Can I allow myself to be meditative and intuitively balanced in my thinking?	Counselor, Pillar of Society
5th Body: Physical Body	This is the body of the teacher. The physical body is a vehicle for divine knowledge.	Teacher/ Balance	Can I allow myself to use my physical capacity to share knowledge with others?	Teacher, Leader
6th Body: Arc Line Body	This is the body of personal integrity. It is the energy we send out into the world and our ability to manifest.	Power of Prayer	Can I allow myself to be truthful and act with integrity?	Dedicated to family, community, humanity
7th Body: Auric Body	This is the body of surrounding light. It is protective and uplifting. The arc line is a part of the aura.	Platform of Elevation	Will I allow myself to uplift and nurture myself and others?	Philosopher, one who elevates through presence and voice

8th Body: Pranic Body	This is the body that deals with energy that comes in through the breath.	Finite to Infinite	Will I allow myself to be fearless?	Healer, Business Person
9th Body: Subtle Body	This body is our sensitive depth of understanding	Mastery or Mystery	Will I allow myself to be subtle and calm?	Advisor, Recorder, Cosmic Caretaker
10th Body: Radiant Body	This body deals with royalty and regalness. It is carried by the aura.	All or Nothing/ Creative Consciousness	Will I allow myself to be creative and courageous?	King or Queen
11th Embodiment: Parallel Unisonness, Non-Duality, Wahe Guru Experience	This embodiment deals with Universal Truth and Divine Perception.	Infinity	Will I allow my identity to be Sat Nam and my experience to be Wahe Guru?	Vibratory Frequency of Siri Guru Granth Sahib and the Sound of Wahe Guru

Epilogue

The more we know about the science of human physiology, the more we can optimize our human existence in every dimension. The development of a peaceful and productive society and the opportunity to live in a spiritually fulfilling manner is a reality.

We live in a time of rapid transition. Adaptation to environmental changes has become a skill and an attribute that is essential for the survival of the individual as well as the human species.

Kundalini Yoga is a technology and a practice that aids the individual in the creation of optimal levels of awareness and creative consciousness. Every experience an individual has, contributes to the collective consciousness of that individual. Skills acquired during a life of Kundalini Yoga and Meditation change brain structures and behavior and contribute to a culture of compassion.

It is my hope that this work will answer some questions regarding physiology, anatomy and Kundalini Yoga. Knowing the mechanisms of Kundalini Yoga can enhance a practice by giving the practitioner insights into how yoga works, as well as aid in the evolution of the subject.

A special Thank You to Yogi Bhajan, without whom this work would not be possible. He gave us all the inspiration to exert the perspiration that created this book.

May this book inspire those who read it to pursue a Happy, Healthy & Holy Life!

May you be a beacon of inspiration and a forklift for all of humanity.

Sat Nam
&
Wahe Guru!

References

Anthony, Catherine Parker, R.N., B.A., M.S., Textbook of Anatomy and Physiology; 7th edition, 1967, The C. V. Mosby Company, Saint Louis

Anthony, Catherine Parker, R.N., B.A., M.S., Textbook of Anatomy and Physiology; 10th edition, 1979, The C. V. Mosby Company, Saint Louis

Bhajan, Yogiji, The Ancient Art of Self Healing; 1979, edited by Dr. Siri Amar Singh Khalsa, D.C.; 1982, Silverstreak Press, Ltd.

Bogduk, Nikolai, BSc(Med) MBBS, Ph.D.; Twoney, Lance T., BappSc(Hons), Ph.D., Clinical Anatomy of the Lumbar Spine; 1987, Churchill Livingstone, Melbourne, Edinburgh, London & New York

Clemente, Carmine D., Anatomy, A Regional Atlas of the Human Body; 3rd edition, Urban & Schwarzenberg

Crouch, James E., Essential Human Anatomy – A Test Atlas; 1982, Lea and Febiger, Philidelphia

Goldberg, Stephen, M.D., Clinical Anatomy Made Ridiculously Simple; 1988, MedMaster, Inc., Miami, Florida

Gates, Douglas, D.C., Correlative Spinal Anatomy; 1984, Tri-State Press, Long Creek, South Carolina

Gray, Henry, F.R.S., Gray's Anatomy, The Classic Collection Edition; Anatomy – descriptive and Surgical; 15th edition, Bounty Books, New York

Guyton, Arthur C., M.D., Structure and Function of the Nervous System; 1972, W. B. Saunders Company, Philadelphia

Guyton, Arthur C., M.D., Textbook of Medical Physiology; 7th edition, W. B. Saunders Company, Philadelphia

Hoppenfeld, Stanley, M.D., Physical Examination of the Spine and Extremities; 1976, Appleton-Century-Crofts, a division of Prentice-Hall

Kapandji, I. A., The Physiology of the Joints, Volume 3, The Trunk and the Vertebral Column; 1974, 2nd edition, Churchill Livingstone, Edinburgh, London, New York

Kapit, Wynn; Elson, Lawrence M., Ph.D., The Anatomy Coloring Book; 1977, Harper and Row, New York

Kapit, Wynn; Macey, Robert I.; Meisami, Esmail, The Physiology Coloring Book; 1987, Harper and Row, New York

Kierman, John A., M.B., Ch.B., Ph.D., D.Sc., Introduction to Human Neuroscience; 1987, J. B. Lippincott Company, Philadelphia

Lad, Dr. Vasant, Ayurveda – The Science of Self Healing; 1984, 2nd edition 1985, Lotus Light Publishing

Marieb, Elaine N., R.N., Ph.D., Human Anatomy and Physiology; 1989, The Benjamin/Cummings Publishing Company, Inc.

Netter, Frank, M.D., The CIBA Collection of Medical Illustrations, Volume I, Nervous System, Part I, Anatomy and Physiology; 1986, 2nd printing, Commissioned and Published by CIBA

Richard, Raymond, D.O., Osteopathic Lesions of the Sacrum – Physio-pathology and Correction Techniques; 1978, Thorsons Publishing Group, Wellingborough, New York

Stoddard, Alan, M.B., B.S., D.O., Manual of Osteopathic Practice; 1983, 2nd edition, Hutchinson, London, Melbourne, Sydney, Auckland, Johannesburg

--- Osteopathy in the Cranial Field; Sutherland Cranial Teaching Foundation, 1976, 3rd edition, The Journal Printing Company, Kirksville, Missouri

--- Mind and Brain – Readings from Scientific American; 1993, W. H. Freeman and Company, New York

About Yogi Bhajan

**"If you want to learn something, read about it.
If you want to understand something, write about it.
If you want to master something, teach it."
-Yogi Bhajan**

Yogi Bhajan, Master of Kundalini Yoga, was the first to publicly teach Kundalini Yoga. He arrived in the US in 1969 and founded the Healthy, Happy, Holy Organization (3HO) based on his first principle: "Happiness is your birthright." Before this, Kundalini Yoga was never taught outside of the old tradition of secrecy.

With the ever-repeated goal to inspire teachers and leaders rather than gather students, Yogi Bhajan travelled internationally teaching Kundalini Yoga, the yoga of awareness.
Today, Kundalini Yoga is taught in many countries around the world. Kundalini Yoga is a method by which one can achieve the scared purpose of your life. It is a practice that is effective, comprehensive, and "do-able".

About the Author

Dr Hari Simran Singh Khalsa, D.C. (also known as DrYogi) has been teaching Kundalini Yoga and Mediation for over 30 years, and has been practicing Chiropractic since 1993. He is a Senior Teacher and a Lead Teacher Trainer for KRI Kundalini Yoga Aquarian Teacher Training, Levels 1 and 2.

He finished his undergraduate studies in Arizona during the 1980s and moved to the San Francisco Bay Area to attend Life Chiropractic College West. In 1993 he graduated and received the LCCW Clinical Excellence Citation Award. He returned to Arizona where he started his Chiropractic practice in Phoenix. In 2004 Dr Khalsa moved back to the Bay Area where he has a private practice and is a faculty member at Life Chiropractic College West. He is a member of the Khalsa Chiropractic Association and the International Chiropractic Association.

DrYogi has done significant work bringing the technology of Kundalini Yoga and Meditation to the communities that need it, and has also been actively involved in the yoga community in various leadership roles. For many years, he taught yoga and meditation in prisons. He was involved with the Arizona Interfaith Movement, and the Arizona Yoga Association. He was the president of 3HO of Arizona (the regional branch of 3HO (the global Kundalini Yoga organization) for over 10 years. He is a former member of the 3HO Int'l Board of Directors and currently on the KRI (Kundalini Research Institute) Advisory Board. Dr Khalsa is a Co-founder of The Elevation Institute.

He is a registered yoga teacher with IKYTA (International Kundalini Yoga Teachers Association) and Yoga Alliance at the 500-hour level.

His teaching credentials include having taught in the Master's Touch Program with Yogi Bhajan, the Self Mastery program at the Omega Institute and various yoga conferences. Dr Khalsa participated in designing the Level 1, 2 and 3 KRI Teacher Training curriculums. He travels internationally teaching in KRI Aquarian Teacher Training Programs.
In addition to teaching Kundalini Yoga, Dr Khalsa also plays the gong and has recorded 2 Gong CDs, Therapeutic Gong, Vol I & II. Yogi Bhajan personally listened to Dr Khalsa's gong CD to approve it to be granted the KRI Seal of Approval.
Dr Khalsa also travels teaching Sat Nam Rasayan, the ancient healing art of Kundalini Yoga.
Dr Khalsa is grateful for the opportunity to have worked directly with Yogi Bhajan (the teacher who brought Kundalini Yoga to the West in the late 1960's) frequently during his life.

He is married to Sat Rattan Kaur and together they lead KRI Kundalini Yoga Teacher Training programs in the San Francisco area and on Kauai. They have been mentors to hundreds of new Kundalini Yoga teachers and students, and pillars of the Kundalini Yoga community.

On the web: www.DrYogi.com Email Dr Hari Khalsa: DrYogi@DrYogi.com

Index

Symbols

A

B

C

H

I

J

K

N

O

P

T

U

V

W

X

Y

Z

Made in the USA
Lexington, KY
12 May 2014